The Practice of
Ion Chromatography

The Practice of
Ion Chromatography

FRANK C. SMITH, JR.

RICHARD C. CHANG

A Wiley-Interscience Publication

JOHN WILEY & SONS

New York   Chichester   Brisbane   Toronto   Singapore

Library of Congress Cataloging in Publication Data:

Smith, Frank C., 1947–
  The practice of ion chromatography.

  "A Wiley-Interscience publication."
  Includes index.
  1. Ion exchange chromatography.   I. Chang, Richard
Chi-cheng, 1918–   II. Title.

QD79.C453S65 1983   543'.0893   82-23914
ISBN 0-471-05517-4

Printed in the United States of America

10 9 8 7 6 5 4 3 2 1

I dedicate this effort to the three people who always believed I would finish—eventually—Deborah, Kris, and Micki.

Frank C. Smith, Jr.

Preface

Ion chromatography is a term that describes a recent major advance in the analysis of ions. The term ion chromatography has been shortened by common usage to the acronym IC. Ion chromatography successfully combines the powerful separations capabilities of ion-exchange resins with the sensitivity and universality of conductimetric detection. IC is uniquely powerful in the solution of determination problems to which the following points apply:

1. High sensitivity is required.
2. Multiple-ion determinations in a sample are preferred.
3. Matrix problems cause interference in other analytical techniques.
4. Specificity in analyzing similar ions is necessary.

Since the introduction of IC in 1975 by Small, Stevens, and Bauman of the Dow Chemical Company, the unique features of ion chromatography are finding application in a growing number of areas of analytical chemistry. The list of new uses and scientific publications has been expanding at a very rapid pace since the technique was commercialized in the fall of 1975 by the Dionex Corporation, the sole licensee of the Dow Chemical Company. Symposia on ion chromatographic methods and applications have been held at the Pittsburgh Conference on Applied Spectroscopy and Analytical Chemistry, at several Rocky Mountain Analytical Conferences, and at EPA. Much of the information in this book was presented at these meetings. In addition, both of us are former employees of Dionex Corporation.

This book is meant to emphasize areas that are essential to an understanding of the practice of ion chromatography and the routine and nonroutine uses of an ion chromatograph. Chapter 1 begins with a short historic account of how and why the IC technique was developed. A brief description of the commercialization of ion chromatography is included. Chapter 2 describes the instrumentation that existed at the time this text was prepared. Chapter 3 offers detailed descriptions of instrument-operating conditions in various modes of suppressor-type IC. Chapter 4, which deals with nonsuppressed ion chromatography, covers these important extensions of the ion chromatographic method. Applications, methods, and column packings used in nonsuppressed IC are discussed.

The major sections on practical IC begin with Chapter 5, which describes the characteristics of the heart of the ion chromatograph—the specialized carbon-based ion-exchange resins used as column packings. Chapter 6 explains IC applications in detail. Chapters 7 and 8 cover IC methods development and advanced IC problem solving as well as expected future developments.

These descriptions of existing routine and advanced IC methods should promote a thorough working knowledge of the practice of IC and provide a perspective of its analytical potential. After finishing this book it is hoped that scientists who have spent the effort to read it will have become familiar with the principles and practices of ion chromatography.

FRANK C. SMITH, JR.
RICHARD C. CHANG

Fremont, California
Mount Vernon, Indiana
March 1983

Acknowledgments

(F.C.S.) should like to express my sincere gratitude to Ed Johnson for his critical comments and red pencil work. Both of us have kind words for Wiley-Interscience for waiting so patiently for the final text.

F.C.S.
R.C.C.

Contents

The Practice of
Ion Chromatography

Ion Chromatography History

The ion chromatograph was developed to solve several specific analytical problems in aqueous systems. Classical methods of separation and quantitative determination of anions and cations in aqueous solutions generally require different reagents and methods for each type of ion. The sample matrix itself sometimes creates severe interference problems, and selectivity between chemically similar species has often been difficult, if not impossible, to obtain. Thus determination of trace quantities of ions in the presence of large amounts of similar ions requires special analytical methods. Finally, ion analysis at ultratrace levels [defined here as determinations at less than 10 µg/L or parts per billion (ppb)] is very difficult and tedious.

The separation of ionic species naturally falls to ion-exchange resins. These resin materials have been used for some time to remove ionic impurities, to concentrate one or more species, to catalyze chemical reactions, and to separate ions by column chromatography. Numerous ion-separation schemes were developed, but no simple instrument that incorporated any of these methods had been commerically marketed.

In-line detectors, which offer a more rapid reproducible analysis, were just beginning to appear when IC was conceived. At that time several detection systems were in general use in liquid chromatographic analysis but were inadequate for low-level analysis of many common ions. Photometric detectors, widely used in HPLC, require that separated species absorb light at one wavelength while the eluent is nonabsorbing at that wavelength. Many common ions absorb light only at low wavelengths. Some may be converted to chromophores (the basis for classical wet chemical/spectrophotometric methods), but each requires different reagents, reaction conditions, and wavelengths for analysis. These restrictions made single and variable wavelength devices insufficiently general for use as ion detectors. Some dedicated ion analyses, based on photometric methods and typified by amino-acid analysis with ninhydrin and o-phthaldehyde to develop color or fluorescense, had been developed. Refractive index detectors are difficult to operate at the high sensitivities needed for trace ion analyses. Electrochemical detectors, which require that the ionic species be oxidizable or reducible, were only in the embryonic stages of development.

Because ions in solution exhibit conductivity to some degree, a good universal ion detector can be based on conductivity measurements. One major problem prevented widespread application of this device. The background conductivity from the eluent was high enough to make trace analysis of the separated ions difficult. Other problems involved cell design and temperature compensation.

Suppressor-type IC was conceived to eliminate the problem of high background conductivity and to develop conductivity detection for routine ion analysis. Dr. William Bauman of Dow suggested the key idea behind IC. He proposed that a second resin be used in series with a separating resin to lower background conductance. Hamish Small initiated the first practical IC work. He discovered that the second resin bed and the conductivity cell would act together as detector and packed a complete IC column set into a single column. A conventional quaternary ammonium anion exchange resin in hydroxide form was packed into the lower half of the column. This was the "suppressor resin." The upper part of the column, packed with low-capacity, surface-sulfonated styrene-divinylbenzene resin, served as the separating resin. The other parts of the prototype IC, which included an injection valve, a reciprocating liquid pump, and a homemade conductivity-measuring device, were assembled for this trial from standard laboratory equipment. The $0.01M$ HCl eluent successfully and repeatedly separated sodium and potassium ions into sharp peaks.

One difficulty existed for this system. As the eluent was converted from high conductivity HCl to low conductivity H_2O, the suppressor part of the resin was converted from the active hydroxide form to the non-suppressing chloride form. Eventually the entire resin bed was completely converted to the chloride form. At this point this dual function column was no longer able to suppress the HCl eluent. The solution to this problem was simply to provide different columns for ion separation and eluent suppression. Regeneration of the suppressor resin with NaOH to convert it from the depleted chloride form to the active hydroxide form required another pump, a timer, and some special valving for rapid and convenient operation.

Shortly after the initial successful experiments the first IC prototype instrument (Figure 1.1) was assembled by Small and Stevens. It was a rack-mounted chromatograph built from commercially available components. The liquid switching and regeneration were manually operated. The results of numerous feasibility experiments were published in late 1975 in "Novel Ion Exchange Chromatographic Method Using Conductimetric Detection." This paper was the first public disclosure of the orig-

Figure 1.1.   Original Dow prototype ion chromatograph. Courtesy of Dow Chemical Corp.

inal ion chromatographic method.[1] Several patents on suppressor-type ion chromatography were also disclosed.

In early 1975 a sole and exclusive license to manufacture and sell ion chromatographs was granted to the Dionex Corporation. Several relatively compact, prototype benchtop ion chromatographs based on the design criteria of the Dow prototype were built. In August 1975 the first commercially available IC was introduced at the National American Chemical Society meeting in Chicago. The method was termed "ion chromatography" rather than "conductimetric chromatography," as proposed by Small et al. Many of the early IC units were used in air-pollution analysis, the first market developed for IC. Other areas in which IC quickly became useful included water-pollution analysis, energy and power production related analysis, quality control, brine and caustic analysis, and trace analysis of all types. Figures 1.2 and 1.3 are standard anion and cation chromatograms obtained on suppressor-type IC.

Several major symposia have been held on ion chromatography. The United States Environmental Protection Agency sponsored two major meetings on "Ion Chromatographic Analysis of Environmental Pollu-

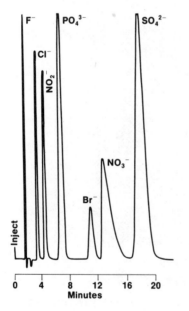
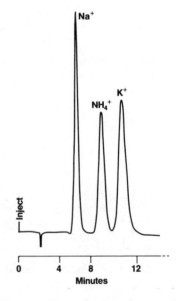

Figure 1.2. Typical inorganic anion separation. Copyright by Dionex Corp. and reprinted by permission of copyright owner.

Figure 1.3. Determination of monovalent cations. Copyright by Dionex Corp. and reprinted by permission of copyright owner.

tants." The first was held in October 1977, the second, one year later. At the first symposium 15 papers on various IC applications were presented in a three-day period; at the second more than 30 papers were heard in a four-day program. Topics at both meetings covered numerous IC applications and advanced methods, and several papers on basic research to extend the capabilities of IC, including the introduction of coupled chromatography, were presented. The proceedings of these meetings have been compiled and published in two books.[2,3] Recognition of the uniqueness of IC as an analytical tool was given further emphasis at a symposium on IC held at the 1979 Pittsburgh Conference on Applied Analytical Chemistry and by major symposia at the Rocky Mountain Analytical Conferences in 1980, 1981 and 1982. A few users' meetings also served as forums for the discussion of applications and problems in ion chromatography.

Several different approaches in the technique of chromatographic ion determinations have recently been introduced, some of which are directly related to the original IC concept of separating ions with resins and have conductivity as the basis of detection. Other methods use different detectors, generally without need for a suppressor column. Electrochemical,[4] UV-visible,[5,6] atomic absorption,[7] and resistivity detectors[8] have been used to detect ions after separation by pellicular or fully porous ion-exchange resins. These methods are complimentary to suppressor-type IC and IC columns and instrumentation have been combined in many of these applications.

Two alternative methods of ion analysis using ion-exchange separation and conductivity detection now are available. In both methods no suppressor column is needed. Instead, moderate conductivity eluents are used to elute a variety of ions. One technique is a variation of conventional HPLC, in which silica-based column packings provide ion separations. Figures 1.4 and 1.5 show typical standard chromatograms obtained by this method. In a second similar approach, specially synthesized macroporous styrene–divinylbenzene resins with low capacities are coupled with moderate-conductivity mono- or polyvalent eluting ions.[9–12] Figures 1.6 and 1.7 are typical chromatograms obtained by this method.

The number of areas in which IC has been applied has grown rapidly and will continue to expand. Because most emphasis has been placed on anion separations, the greatest number of species currently separable by IC are anionic. Similar increases for cationic species should occur in the future. The term ion chromatography is now most popularly applied to

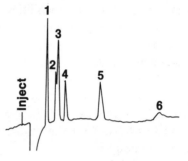

Column: VYDAC 302IC4.6
Detector: VYDAC 6000 CD—Conductivity Detector
Solvent: 4mM Phthalic acid, pH 4.5 c̄ NaBorate
Flow Rate: 2ml/min
Injection: 200 microliters of:
    1. Chloride—10ppm
    2. Nitrite—10ppm
    3. Bromide—10ppm
    4. Nitrate—10ppm
    5. Sulfate—10ppm
    6. Thiosulfate—20ppm
    7. Cyanide—20ppm

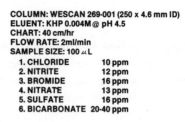

COLUMN: WESCAN 269-001 (250 x 4.6 mm ID)
ELUENT: KHP 0.004M @ pH 4.5
CHART: 40 cm/hr
FLOW RATE: 2ml/min
SAMPLE SIZE: 100 μL
    1. CHLORIDE   10 ppm
    2. NITRITE   12 ppm
    3. BROMIDE   16 ppm
    4. NITRATE   13 ppm
    5. SULFATE   16 ppm
    6. BICARBONATE   20-40 ppm

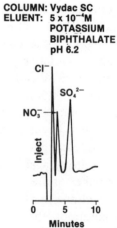

Figure 1.5. Single-column separation of anions. Courtesy of Wescan Corp.

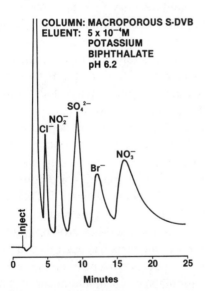

14 min.

Figure 1.4. Nonsuppressed separation of common anions. Courtesy of Separations Group.

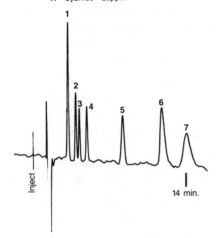

COLUMN: MACROPOROUS S-DVB
ELUENT: 5 x 10⁻⁴M POTASSIUM BIPHTHALATE pH 6.2

Figure 1.6. Macroporous S-DVB separation of anions. Reprinted from the *Journal of Chromatography* with permission.

COLUMN: Vydac SC
ELUENT: 5 x 10⁻⁴M POTASSIUM BIPHTHALATE pH 6.2

Figure 1.7. Rapid separation of anions by nonsuppressed IC. Courtesy of Wescan Corp.

ion-exchange separations of simple ions in which detection is based on conductivity. As other separations are developed and put into common use and as other detectors become more routine in ion analysis the term IC will undoubtedly be expanded to describe any chromatographic method of ion analysis of both simple and complex ions.

NOTES

1. H. Small, T. S. Stevens, and W. C. Bauman, "Novel Ion Exchange Chromatographic Method Using Conductimetric Detection," *Anal. Chem.*, **47**(11), 1801–1809 (1975).
2. E. Sawicki, J. D. Mulik, and E. Wittgenstein, Eds., *Ion Chromatographic Analysis of Environmental Pollutants*, Vol. 1, Ann Arbor Science, Ann Arbor, Michigan, 1978, 210 pp.
3. J. D. Mulik and E. Sawicki, Eds., *Ion Chromatographic Analysis of Environmental Pollutants*, Vol. 2, Ann Arbor Science, Ann Arbor, Michigan, 1979, 435 pp.
4. J. E. Girard, "Ion Chromatography with Coulometric Detection for the Determination of Inorganic Ions," *Anal. Chem.*, **51**(7), 836–839, (1979).
5. R. J. Williams, "Application of Ion Chromatography to the Analysis of Organic Sulfur Compounds," presented at the Rocky Mountain Conference on Analytical Chemistry, Denver, Colorado, August 1980.
6. E. Cathers and E. L. Johnson, "Applications of Simultaneous Detection in IC," presented at the Rocky Mountain Conference, Symposium on Ion Chromatography, Denver, Colorado, August 1981.
7. G. Colvos, N. H. Hester, G. R. Ricci, and L. S. Shepard, "Ion Chromatography with Atomic Absorption Spectrometric Detection for Determination of Organic and Inorganic Arsenic Species," *Anal. Chem.*, **53**(4), 610–613 (1981).
8. R. K. Pinschmidt, Jr., "Ion Chromatographic Analysis of Weak Acids Using Resistivity Detection," in J. D. Mulik and E. Sawicki, Eds., *Ion Chromatographic Analysis of Environmental Pollutants*, Vol. 2, Ann Arbor Science, Ann Arbor, Michigan, 1979, pp. 41–50.
9. D. T. Gjerde, "Ion Chromatography with Low Capacity Resins and Low Conductivity Eluents," Doctoral Thesis, Iowa State University, Ames, Iowa, *Diss. Abs. Int. B*, **41**(6), 2166 (1980).

10. D. T. Gjerde, J. S. Fritz, and G. Schmuckler, "Anion Chromatography with Low-Conductivity Eluents," *J. Chromatogr.*, **186**, 537–547 (1979).

11. D. T. Gjerde, G. Schmuckler, and J. S. Fritz, "Anion Chromatography with Low-Conductivity Eluents II," *J. Chromatogr.*, **187**, 35–45 (1980).

12. J. S. Fritz, D. T. Gjerde, and R. M. Becker, "Cation Chromatography Using a Conductivity Detector," *Anal. Chem.*, **52**, 1519 (1980).

Existing Instrumentation

Since there are several different approaches to ion chromatography, it naturally follows that several variations in instrumentation have developed. These instruments range from liquid chromatographs (LC), which have been modified by an add-on column/detector package, all the way to dedicated, fully automated ion chromatographs. Despite some significant differences in detail, instruments for any type of IC have numerous features in common. Naturally the content of this chapter has a somewhat limited lifetime because instrument manufacturers are continually updating and improving their products.

Most IC analyses are now performed on suppressor-type instruments. Some unique advantages to suppressor columns are claimed[1] although, in the authors' opinion, nonsuppressed methods are fully capable of performing many of the same analyses that can be done on suppressed systems. Conventional HPLC components can readily be assembled to do nonsuppressed IC. Modified LC systems and nonsuppressed IC instruments may be used for some types of suppressed IC, but the use of these instruments in this fashion may be in conflict with the patents on eluent suppression. Appropriate column sets to perform suppressed IC are not generally available except to purchasers of Dionex equipment . Moreover, these instruments are not set up for convenient suppressor regeneration and they must be substantially modified to do coupled chromatography. The instruments designed for suppressor-type IC can also be used for nonsuppressed ion analysis with only minor changes. Because the emphasis in this book is on practical IC and because suppressor-type instruments are used in most analyses, they are covered in more detail. When applicable, significant differences between existing suppressor-type and nonsuppressed IC instrumentation are explained.

The only suppressor-type ion chromatograph is marketed by Dionex Corporation (Sunnyvale, California). Dedicated ion chromatograph systems for nonsuppressed IC have recently been introduced by Wescan Instruments (Santa Clara, California) and Hewlett-Packard Instruments (Palo Alto, California). Wescan, Tracor Instruments (Austin, Texas), and The Separations Group (Hesperia, California) offer conductivity detectors for use with HPLC equipment. The columns in most nonsuppressed instruments are packed with silica-based ion exchangers developed by The Separations Group and marketed under the Vydac® trademark. The number of manufacturers of IC equipment or of equipment to enable the users of an LC to do IC analyses should increase in the near future.

The components of an ion chromatograph perform functions identical to those of any liquid chromatograph. As shown in Figure 2.1 there are five major components in a suppressor-type ion chromatograph:

1. Eluent pump and liquid containers.
2. Sample injection valve.
3. Ion-exchange separating column.
4. Suppressor column coupled to conductivity detector, meter, and output device.
5. Regenerating pump with electronic timer and control.

Each of these components is discussed separately.

The eluent pumping system for the ion chromatograph consists of eluent containers, liquid transfer lines, eluent selection valves, and a pump. Obviously these components pump the eluents, thus pushing the eluting ions through the columns and detector. The eluents can be stored in collapsible polyethylene containers or glass flasks, which are fairly inert chemically and have the added advantage of minimizing interaction of the eluent solution with the atmosphere. Both liquid containers allow some exposure to light unless wrapped in foil, heavy paper, or a cardboard box.

The prototype of a Dow ion chromatograph had Teflon® tubing for liquid lines. This chemically inert material does not interact with corrosive eluents like HCl and NaOH. Until very recently, Teflon tubing was still

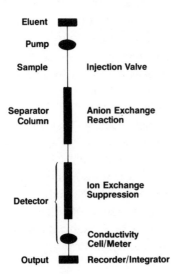

Figure 2.1. Basic components of an IC. Copyright by Dionex Corp. and reprinted by permission of copyright owner.

used in most suppressor-type instruments. New suppressor-type ion chromatographs now contain Tefzel® tubing for the smaller diameter high-pressure lines and Teflon tubing for the large bore, low-pressure pump supply lines. The burst pressure rating of the Tefzel small-diameter tubing is about 2000 to 3000 psig, which raises the maximum system operating pressure above the 900 to 1000 psig limit for Teflon. Both types of plastic tubing are interfaced to the valves and columns by a patented gripper system. This method had previously required hot flaring of the Teflon tubing.

Nonsuppressed IC instruments use small diameter stainless steel for high-pressure tubing and Teflon tubing for pump inlet lines. Because the eluents needed for these chromatographic separations are quite mild, corrosion of the stainless steel does not appear to be a problem. These systems can operate at high pressures (approximately 2000 to 3000 psig.). Conventional HPLC instruments modified to do IC also have stainless steel tubing. If off-line regeneration capability is established, it is quite straightforward to couple a suppressor column to the separator column and to use the HPLC to do suppressor-type ion chromatography. The halide-based eluents used in some types of suppressed IC will cause corrosion problems in this tubing. Care *must* be taken to flush the pump with deionized water after each use. Otherwise, the pump seals will need to be replaced more frequently than normal. Other types of suppressed IC (standard anion, for example) can be done on this equipment if appropriate column sets and sublicenses can be obtained. The use of equipment in this manner may infringe on patents on eluent suppression.

Several plastic valves provide eluent choices for step gradient changes, to select a high-concentration "washing" solution to remove strongly retained species from the separator column, or to provide for washing column sets in water. In one convenient arrangement two or three valves are coupled in a master-slave arrangement to provide eluent selection. Another arrangement allows direct selection of six eluents.

The eluent pump must produce reliable "constant" delivery of eluents to do routine reproducible species identification and quantitative determinations. Typical eluent flow rates for ion chromatographic analyses range from 0.7 to 4 mL/min. The pump previously used in most IC instruments was a single-piston reciprocating unit which was inexpensive and provided adequate flow reproducibility and regulation. Flow rate is adjusted by changing the travel of the piston with a variable stop. Two sets of check valves ensure that liquid flow is unidirectional. A bubble trap is used to keep the pump head from filling with air (which causes loss of liquid prime). The high degree of mechanical flexing of the resin caused by the alternating filling of the chamber and then pushing the eluent through the high-pressure tubing does not appear to damage the styrene/divinylbenzene or silica resins physically. Baseline noise is a bothersome problem with this type of pump unless pulse dampers are used. The pulsing is somewhat reduced by a mechanical pressure gauge which also allows a continuous check of the liquid "tightness" of the rest of the high-pressure components. Alternatively, a pulse damper designed for liquid chromatographs can be used to achieve less noisy baselines.

Pulseless pumps common to HPLC seem to provide major advantages in maintaining constant flow. This is especially important in a flow-sensitive system at low flow rates, such as an ICE (ion chromatography exclusion) column set. The present generation of IC instruments includes these pumps, with some made from plastic.

One of the most important high-pressure components is the sample injection valve. Plastic valves made of chemically inert materials are used in suppressor-type IC instruments. These valves include Kel-F® sliders, Teflon port faces, and polypropylene valve bodies and air actuators. A fixed-volume Teflon loop holds the sample and the eluent stream is diverted through this loop to "inject" the sample. The valve is air operated to ensure smooth, reliable actuation and to make automation easy. In the nonsuppressed instruments rotary type stainless-steel valve commonly used in HPLC instruments is used to inject samples. Again, a fixed volume loop holds the sample and the eluent stream is diverted through the loop to "inject" the sample.

For ultratrace ion concentrations loop injection may not provide sufficient sensitivity. In these cases sample injection is done with a concentrator column used in place of the sample loop. Large measured volumes of sample (typically 5 to 50 mL) are pumped through the concentrator column, which strips out the ions of interest. The sample is injected by diverting the eluent through the concentrator column instead of the sample loop. Sufficient eluent must pass through the concentrator to elute the ions of interest. Concentrator columns are also used in coupled chromatography applications in which concentrators trap selected species that are eluted and detected in the first system [ion exclusion (IE), ion chromatography exclusion (ICE), or ion chromatography (IC)] for subsequent analysis on a second column set. Figure 2.2 is a basic flow schematic of an instrument capable of doing coupled chromatography. This instrument

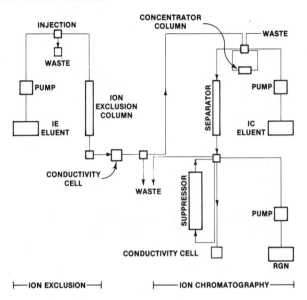

Figure 2.2. Coupled chromatography instrument schematic. Copyright by Dionex Corp. and reprinted by permission of copyright owner.

is really two basic pumping systems and two detectors in a single unit. Some additional valves and tubing are used to increase the unit's versatility in coupled chromatography.

After being "injected" the samples are pushed as a liquid plug onto the top of a precolumn. This column is used in series before the separator column and is filled with resin very similar to the separating column resin. The primary function of the precolumn is to protect the more expensive separator column by absorbing organic materials, trapping particulates from the sample and eluent, and retaining polyvalent ions and surface fouling species. Once the sample ions have been exchanged onto the top of the precolumn the eluent flow begins the differential migration of the ions from the precolumn. After the eluent ions begin to pass through the precolumn the weakly retained ions begin to move quickly onto the separator column. More strongly held species remain on the precolumn longer before moving onto the separator column. Both the precolumn and separator column use pellicular packings to obtain optimum separation with adequately short analysis times. Detailed descriptions of column packings are provided in Chapter 5.

After the separated species emerge from the separator column they pass into the detector. The suppressor-type IC has a unique detector in

Table 2.1   Suppressor Column Reactions

Anion Suppressor

Equation 1. $(H^+\text{-resin}) + NaOH \rightleftharpoons (Na^+\text{-resin}) + H_2O$
           resin phase   eluent   resin phase   column effluent

Equation 2. $(H^+\text{-resin}) + NaHCO_3 \rightleftharpoons (Na^+\text{-resin}) + H_2CO_3$

Equation 3. $2(H^+\text{-resin}) + Na_2CO_3 \rightleftharpoons 2(Na^+\text{-resin}) + H_2CO_3$

Conductivity Enhancement

Equation 4. $NaCl + K_2SO_4 + 3(H^+\text{-resin}) \rightleftharpoons$
$$HCl + H_2SO_4 + (Na^+\text{-resin}) + 2(K^+\text{-resin})$$

Cation Suppressor

Equation 5. $(OH^-\text{-resin}) + HNO_3 \rightleftharpoons (NO_3^-\text{-resin}) + H_2O$

Equation 6. $m\text{-PDA}\cdot 2HCl^* + 2(OH^-\text{-resin}) \rightleftharpoons$
$$2(Cl^-\text{-resin}) + m\text{-PDA} + 2H_2O$$

Conductivity Enhancement

Equation 7. $NaCl + KBr + (OH^-\text{-resin}) \rightleftharpoons$
$$NaOH + KOH + (Cl^-\text{-resin}) + (Br^-\text{-resin})$$

ICE Suppression

Equation 8. $(Ag^+\text{-resin}) + HCl \rightleftharpoons (H^+\text{-resin}) + AgCl\downarrow$

ICE Enhancement (in post-suppressor column)

Equation 9. $(H^+\text{-resin}) + HCOO^-Ag^+ \rightleftharpoons (Ag^+\text{-resin}) + HCOO^-H^+$

---

* $m\text{-PDA}\cdot 2HCl$ = meta-phenylenediamine dihydrochloride.

that a second ion-exchange column is an integral part of the detector. This column, called the suppressor column, is used in series between the separating column and flow-through conductivity cell. The suppressor column resins react with the eluent (Table 2.1) to reduce the background conductivity by several orders of magnitude. In addition, all the sample ions are converted to a single, more conductive form. Sample ions are typically detected in the H^+ or OH^- form (i.e., H^+Cl^- for Cl^- analysis and Na^+OH^- for Na^+ analysis). These species are highly conductive and easily detectable on the low conductivity background.

The suppressor resin is eventually completely converted to a nonsuppressing form. The high background conductivity makes trace analysis almost impossible. The column can be regenerated to the active form by a strong electrolyte solution. A separate regeneration pump and valving circuit are electronically controlled to flush the suppressor column, first with regenerating solution and then with water. The pump is automatically shut off at the end of the cycle. Typical regenerants are $2M$ H_2SO_4 for anion suppressors and $0.5M$ NaOH for cation suppressors. These strong solutions require that liquid tubing and suppressor-column materials be inert. Because of the valving arrangement in suppressor-type IC, one suppressor can be used in-line for eluent suppression while another column is regenerated off-line. Alternation of these two columns results in IC operation with almost no downtime for regeneration.

Alternatively, off-line regeneration of the suppressor is possible if another pump is available. This simply requires two suppressor columns, one that is being regenerated and a second installed for analytical use. This allows a conventional HPLC instrument with a conductivity detector to be used to do suppressor-type IC, although this will infringe on the patents on eluent suppression.

One suppressor column innovation has recently been developed. The resin in the suppressor column is replaced by a filled hollow fiber ion exchange membrane.[2] This membrane acts exactly like the suppressor resin in that ions are exchanged from the membrane for ions in the eluent stream. The innovation is that the membrane is continuously regenerated by a gravity-fed flow of low-concentration H_2SO_4 which continuously replaces the ions that are exchanged onto the fiber with ions from the regenerant. Thus separate regeneration steps are eliminated. The advantages of this sytem over a suppressor column appear to be continuous operation without regeneration and the use of lower strength regenerants. These columns also reduce the interaction of weak acids with the suppressor and minimize the effect of the "water dip" in standard anion analysis. Disadvantages appear to be related to slight initial difficulties in installation, although other problems may appear as the system becomes more widely accepted. This suppression system was not in widespread use at the time this text was prepared, so disadvantages could not easily be identified.

In some separation systems the suppressor columns are difficult to regenerate. These systems have made effective use of disposable suppressor columns. One excellent example is the ICE system in which the suppressor column acts by precipitating the Cl^- ion from the HCl eluent with Ag^+ (from the suppressor) and by exchanging the H^+ onto the vacant resin site. The silver chloride precipitate is held in the resin network. As the AgCl builds up, system pressure rises. If the pressure rise is allowed to continue, maximum system operating pressure limits can be exceeded and can result in a component failure. Also, the single-piston reciprocating pumps do not seem to operate as reproducibly as desired at low flow rates when the system back pressure changes. The regeneration process for an ICE suppressor involves dissolving the AgCl with NH_4OH/NH_4Cl, washing extensively with water, resilvering the resin with $AgNO_3$, and washing again with water. This is a lengthly process and quickly led to the development of disposable plastic column refills. With these columns the expended portion of the suppressor is cut off daily and the end fittings are moved onto the remaining length of column. When the entire length of column is expended, the resin-filled tube is replaced. A second suppressor-type column, the ICE post-suppressor, converts the separated salts from the Ag^+ form to the more conductive, more reproducible H^+ form. This column is not routinely expended and will often last two to three months or more. The authors have encountered some difficulties in repairing leaking columns and in adding resin to eliminate headspace. Careful adherence to manufacturers' instructions should minimize these difficulties.

In nonsuppressed IC no suppressor column is required. Because this column is not used, the dead volume after the separator column is less. In theory this should lead to more efficient separations, although the increase seems to be small.

Columns used in suppressor-type ion chromatography were initially made of glass. Lower priced plastic columns have been introduced. The performance of these columns is equivalent to or better than that of glass columns and there is no breakage, but some difficulty in repairing leaks has been observed. The plastic columns are packed by conventional LC packing procedures, whereas glass columns, because of the high flow velocities used in these procedures, are not. Typical plastic column internal diameters are 4, 6, and 9 mm; their lengths vary from 50 to 250 mm. These columns are presently sold by Dionex only to owners of their ion chromatographs. Bulk packing materials are unavailable.

Columns in nonsuppressed IC can be glass, plastic, or stainless steel. Most columns now in use are made of stainless steel. The phthalate and benzoate eluents in these separations have pH ranging from 3 to 7; therefore little corrosion is expected. The injection of samples with pH higher than 7 is not advisable because the silica packing will degrade severely. In general, samples in which the pH is greater than 7 must be treated with eluent until the proper pH balance is achieved.

The second part of the detector is a high-sensitivity flow cell and conductivity meter. The cell body is constructed of Kel-F and internal volume is nominally 2 to 6 μL. Electrodes are made of 316 stainless steel. To facilitate temperature compensation a thermistor is placed in the liquid line just after the electrodes. The cell is driven by a high-frequency oscillator from the main circuit board. The cell output drives an amplifier on this same board. Changes in the ionic composition in the cell result in signal changes to the amplifier. After signal processing, which includes temperature compensation, the signal caused by the presence of conductive ions in the cell results in meter and recorder-pen deflection.

Several fully automatic suppressor-type ion chromatographs are presently available from Dionex. An electronic programmer controls the various instrumental parameters that might be changed during an IC determination:

1. Sample injection.
2. Autosampler and sample loop/concentrator column loading pump.
3. Regeneration.
4. Detector range changes.
5. Recorder or data system operation.
6. Eluent step gradients.
7. Column heater.
8. Flow rate gradient.
9. Autozeroing of baseline.

This ion chromatograph is most useful when the samples are similar. Limits on the linearity of the overranging capability of the detector do not allow vastly different ion concentrations to be run in the automatic mode, but it can be used in the manual operation mode for normal suppressor IC work.

The nonsuppressed IC instruments have many of the components of the suppressor-type instruments. The Wescan instrument contains a single-piston reciprocating pump, a mechanical gauge, a heat-sinked and temperature-compensated conductivity cell and meter, and recorders or integrators as output devices. The Hewlett-Packard IC is a conventional HPLC with dual piston pump, microprocessor control, column oven, and other features. The main differences between these nonsuppressed instruments and suppressor-type ion chromatographic instruments lie in the liquid transfer line materials, the column packings, the lack of a regeneration pump and timer in the nonsuppressed systems, and the electronic offset circuit for the background conductivity. This last point is a key difference. Conductivity measurements are quite sensitive to temperature fluctuations if adequate temperature control is not achieved. The higher background in nonsuppressed systems is expected to vary more with small changes in temperature. In nonsuppressed IC the background is lowered by electronic compensation. Signal-to-noise ratios are somewhat lower than those in a suppressor-type column system, largely because of the ionic form of the separated ion when it is detected. The suppressor column converts the separated ions to their most conductive form. Additionally, the silica column packing must be used with eluent and sample pH between 3 and 7; otherwise there is the risk of degrading the packing. In spite of these limitations a single column system is still quite useful in many types of sample where ultimate sensitivity is not needed, where the sample matrix is not too complex, or where automatic operation is not needed.

The ion chromatography instrumentation presently available resembles other liquid chromatographs. The systems have been designed for ruggedness and reliability. Because IC is a specialized type of LC, the developments in these instruments should parallel HPLC instruments. This will include modular as well as integrated component packaging, microprocessor control, and data-handling capabilities. Post column reaction systems and other detectors will be coupled to IC separating columns to expand the range and depth of ion chromatography.

NOTES

1. C. A. Pohl and E. L. Johnson, "Ion Chromatography-State of the Art," *J. Chromatogr. Sci.*, **18**, 442–452 (1980).
2. T. S. Stevens, J. C. Davis, and H. Small, "Hollow Fiber Ion-Exchange Suppressor for Ion Chromatography," *Anal. Chem.*, **53**(9), 1488–1492 (1981).

CHAPTER
3
The Practice of Ion Chromatography

One of the main features of IC is the ability of the ion chromatograph to separate and detect many different anions and cations. For all this versatility a few ions (especially SO_4^{2-} and Cl^-) have been much more frequently analyzed than all the other ions combined. Ion chromatographic methods for analyzing these key ions have been in general use for several years. Instrument-operating conditions for eluting key ions have not changed substantially during this time. This chapter describes the existing common IC practices and is centered around these key ions. Methods development techniques for samples in which the ions of interest have not been analyzed or in which complex analytical problems exist are described in Chapter 7.

The ion chromatograph can be used in a variety of instrument configurations with different column sets and eluents. Despite the numerous possibilities most minor changes in operating parameters do not significantly alter elution profiles. Certain key ions are commonly determined in a wide number of matrices with similar instrumental parameters which have been termed standard operating conditions or simply standard conditions. Five major sets of standard conditions are routinely used:

1. Anion ion chromatography.
2. Cation ion chromatography.
3. Ion exclusion or ion moderated partition.
4. Ion chromatography exclusion.
5. Coupled chromatography.

3.1 CONDITIONS FOR STANDARD CONDITIONS SEPARATIONS

Standard operating conditions have been developed by common use. Although these conditions cannot determine all ions in all matrices, they are versatile enough to allow a large number of ion separations a single injection. When any set of standard conditions is used, no matter whether anionic or cationic species are being determined, several instrument parameters are the same for all analyses.

All standard condition ion separations are done with isocratic (one-eluent) elution. Eluents should be prepared from deionized or distilled water which has been filtered through a 0.2-μ filter. Apparently deionized water is preferred to distilled water. Reagent grade or ultrapure chemicals should be used whenever possible. Typical eluent flow rates occur between 0.7 and 4.5 mL/min, depending on the analytical requirements and system pressure.

The conductivity cell and meter are operated at a cell constant equal to 1.0. The spacing of the electrode is set so that the instrument will respond at 147 μMHO/cm for a standard 0.001M KCl solution. In most IC analyses the conductivity meter is set on a scale between 1 and 100 μMHO/cm. The eluent background conductivity after suppression is generally less than 30 μMHO/cm. A dual pen recorder is used to record output from the IC detector. One recorder channel is set to operate from 0 to 1 V input; the second is set to operate from 0-to-100 mV input. This provides a 10-fold scale expansion for simultaneous measurement of large and small peaks. It also leads to complicated-looking chromatograms.

Normal injection volume is 100 μl, although the actual volume of injection does not need to be accurately measured. Samples that are too high in ion content are manually diluted with water or eluent. Particulates in the samples are removed by filtration. The sample loop is normally loaded from disposable plastic syringes, and 1 to 2 mL of sample are used to flush the transfer line and sample loop. Sample injection is a two-step operation consisting of (a) loading the Teflon loop with sample and (b) diverting the eluent through the loop.

Several columns are typical in any suppressor-type IC analysis. The column configuration (in the order in which the sample ions see them) is generally the following:

1. Precolumn.
2. Separator column.
3. Suppressor column.
4. Post-suppressor column (if required).

These columns are all used in series. Valves and liquid transfer lines are placed between some of the columns; others are directly coupled. For some sets of standard conditions not all the different types of column are required.

Some of the standard separation schemes with more than one set of standard operating conditions have been developed. Each set has individual characteristics which include column type, column diameter, column length, eluent, and eluent flow rate. Table 3.1 is a summary of these parameters. Each set of operating conditions is explained in detail. Column sizes are designated in millimeters by giving the internal diameter first and then the length.

3.2 STANDARD ANION IC CONDITIONS

Standard anion IC conditions can be used in a wide range of anion separations. As a general rule, these conditions apply to species with relative retention from 0.2 to 1.2 times the sulfate ion retention. Two separator column/eluent combinations are used for standard anion conditions. The complexity of the sample, total concentration of ions in the sample, and the ion separation requirements all influence the analyst's choice.

The first and by far the most frequently used anion standard condition utilizes a 3 mm ID × 500 mm glass separator column or a 4 × 250 mm plastic separator column. The eluent contained in these columns is a mixture of $0.003M$ $NaHCO_3$ and $0.0024M$ Na_2CO_3. Flow rates are typically 2.5 to 3.5 mL/min. This eluent/column combination can be used for simple, highly concentrated, or complex samples. Figure 3.1, a typical chromatogram, shows the separation of seven common inorganic anions. Note particularly the complete $Cl^- - NO_2^-$ separation and the adequate sepa-

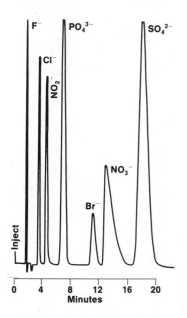

Figure 3.1. Typical inorganic anion separation. Copyright by Dionex Corp. and reprinted by permission of copyright owner.

Table 3.1 Standard Elution Conditions[a]

Instrument Parameter	Suppressor-Type Anion IC				
	Moderately Retained Species			Strongly Retained Species	Weakly Retained Species
	Simple Matrix	Complex Matrix	Rapid Analysis		
Precolumn size (mm)[b]	3 × 150 (g); 4 × 50 (p)	3 × 150 (g); 4 × 50 (p)	4 × 50 mm (p)	3 × 150 (g); 4 × 50 (p)	3 × 150 (g); 4 × 50 (p)
Separator size (mm)[b]	3 × 250 (g); 4 × 100 or 2 ea 4 × 50 (p)	3 × 500 (g); 4 × 250 (p)	Fast run 4 × 250 (p)	3 × 250 (g); 2 ea 4 × 50 (p)	3 × 250 (g); 2 ea 4 × 50 (p)
Suppressor size (mm)[b]	6 × 250 (g) or 3 × 150 (g); 9 × 100 (p) or 6 × 60 (p)	6 × 250 (g) or 3 × 150 (g); 9 × 100 (p) or 6 × 60 (p)	6 × 250 (g) or 3 × 150 (g); 9 × 100 (p) or 6 × 60 (p)	6 × 250 (g) or 3 × 150 (g); 9 × 100 (p) or 6 × 60 (p)	6 × 250 (g) or 3 × 150 (g); 9 × 100 (p) or 6 × 60 (p)
Suppressor active form	H^+	H^+	H^+	H^+	H^+
Most commonly used eluent	$0.003M$ $NaHCO_3$/ $0.0012M$ Na_2CO_3	$0.003M$ $NaHCO_3$/ $0.0024M$ Na_2CO_3	$0.003M$ $NaHCO_3$/ $0.0024M$ Na_2CO_3	$0.006M$ Na_2CO_3	$0.0015M$ $NaHCO_3$
Typical flow rate (mL/min)	3–4	2–3	2–3	2–3	2–3
Approximate analysis time (min)	6–8	15–18	7–8	8–15	12–15
Other eluents used	None	$0.0036M$ Na_2CO_3/ $0.002M$ $NaOH$	None	$0.005M$ NaI[c]	$0.005M$ $Na_2B_4O_7$
Typical ions eluted	F^-, Cl^-, NO_3^-, SO_4^{2-}	F^-, Cl^-, NO_2^-, PO_4^{3-}, NO_3^-, Br^-, SO_4^{2-}	F^-, Cl^-, NO_3^-, PO_4^{3-}, NO_3^-, Br^-, SO_4^{2-}	SCN^-, I^-, $S_2O_3^{2-}$	F^-, Cl^-

Nonsuppressed Anion IC			Non-Ion Exchange IC		Suppressor-Type Cation IC	
Moderately Retained Species	Weakly Retained Species	Strongly Retained Species	Ion Exclusion (IE) or Ion Moderated Partition (IMP)	Ion Chromatography Exclusion (ICE)	Monovalent	Divalent
4.6 × 50 (s)	4.6 × 50 (s)	4.6 × 50 (s)	None	None	3 × 150 (g); 4 × 150 (p)	3 × 150 (g); 4 × 150 (p)
4.6 × 250 (s)	4.6 × 250 (s)	4.6 × 250 (s)	6 × 500 (g); 9 × 150 (p); 7.8 × 300 (s)	9 × 250 (g); 9 × 200 (p)	6 × 250 (g); 6 × 60 (p)	6 × 250 (g); 6 × 60 (p)
None	None	None	None	6 × 150 (p)	9 × 250 (g); 9 × 100 (p)	9 × 250 (g); 9 × 100 (p)
None	None	None	Water or 0.002M H_2SO_4	Ag^+ [c] 0.001 to 0.0001M HCl	OH^- 0.0075M HCl or HNO_3	OH^- 0.0025M mPDA[e] plus 0.0075M $HClO_4$
$(1-5) \times 10^{-3}\ M$ Potassium acid phthalate	0.0033M acetic acid or 0.001M benzoic acid	"Wescan" AN-1 or potassium acid phthalate	0.5–1	0.8–1.0	2–4	2–4
1–2	1–2	1–2	5–30	30	10–20	10–15
10–25	10–15	10–20	Other mineral acids	None	None	None
Potassium benzoate	Aliphatic organic acids	Citrate	Sugars, CO_3^{2-}, organic acids	Organic acids PO_4^{3-}	Aliphatic amines, alkali metals, ammonium	Divalent cations
F^-, Cl^-, NO_3^-, SO_4^{2-}	F^-, Cl^-	SCN^-, I^-, $S_2O_3^{2-}$				

[a] (p) = plastic column; (g) = glass column; (s) = stainless steel column.

[b] Internal diameter dimension × length.

[c] Requires halide suppressor.

[d] 4 × 150 mm post-suppressor column used.

[e] mPDA = meta-phenylenediamine dihydrochloride.

ration of $Br^- - NO_3^-$. Recently, para-cyanophenol has been added to the eluents as an absorption modifier for Br^- and NO_3^- separations. With this additive the Br^- and NO_3^- merge into a single peak which can be moved by careful regulation of the cyanophenol concentration. These same columns, used with $0.003M$ $NaHCO_3/0.0024M$ $Na_2CO_3/0.0005M$ p-cyanophenol, produce excellent separations of six ions in 7 min (Figure 3.2). Another innovation is the introduction of "fast-run" separator columns, which shorten run times significantly without severely compromising the resolution of adjacent peaks.

The time between injections of unmodified eluent had been about 20 to 25 min, the yield, 19 to 24 samples/8-h day (or 130 to 160 ion analyses).

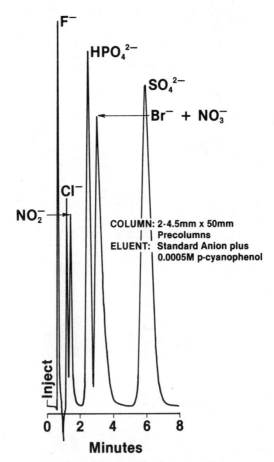

Figure 3.2. Rapid anion separation.

Typically, higher eluent flow rates or use of the fast-run columns will give 8 to 12 min analysis times, although the system pressure should be kept below 750 psig to prevent leaks. This simple change raises sample through-put to 40 to 48 samples/8-h day (280 to 335 ion analyses).

The second anion IC standard condition is used for samples in which the ion composition is simple or a more rapid analysis is required for a few well separated species. A 3×250 mm glass column or two 4×50 mm plastic anion precolumns are used with $0.003M$ NaHCO$_3$/$0.0024M$ Na$_2$CO$_3$ or $0.003M$ NaHCO$_3$/$0.0012M$ Na$_2$CO$_3$. The first eluent fails to provide complete separation for NO$_3^-$ and SO$_4^{2-}$, although it is usually adequate for most analytical purposes. The second eluent gives baseline NO$_3^-$–SO$_4^{2-}$ separation and is normally used at higher eluent flow rates for rapid analysis (Figure 3.3). These conditions will allow 60 to 70 samples to be run in an 8-h day (equivalent to 240 to 280 ion analyses).

The most commonly used anion suppressor column is a 6×250 mm glass column or a 9×100 mm plastic column. These columns provide eluent suppression for 12 to 14 h before regeneration is required. One problem associated with large glass columns is a large water dip that interferes with early eluting peaks like Cl$^-$. The problem does not seem so severe when plastic suppressors are used. Because analysis of trace levels of chloride is an important IC application, smaller suppressor columns (3×50 mm or 3×150 mm glass or 6×60 mm plastic) have been

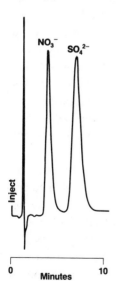

Figure 3.3. NO$_3^-$ and SO$_4^{2-}$ determination (9 min). Copyright by Dionex Corp. and reprinted by permission of copyright owner.

used for these determinations. The small column minimizes the broadness of the water dip, thus allowing the Cl^- peak to be better separated from the dip. A two to three hour period is the typical operating cycle for the smaller suppressors which are most often used in pair, with one column in-line for eluent suppression while the other column is being regenerated. Continuous instrument operation is possible with this operating configuration. Alternatively, the hollow fiber suppressor causes the "water dip" to occur well before the chloride peak.

Anion suppressor columns are regenerated by $2M$ H_2SO_4, then washed in water. Typical regeneration times are 30 min (10-min regenerant wash and 20-min water wash). Reequilibration with eluent requires another 5 to 10 min. Shorter times are used for small suppressors.

3.3 STANDARD CATION IC CONDITIONS

Cation IC standard conditions normally use different separator columns, depending on the ions of interest. For alkali (Group I) metals, ammonia, short-chain aliphatic amines, and some small quaternary ammonium compounds a 6×250 mm glass or a 4×200 mm plastic separator column is used with 0.005 or $0.0075M$ HNO_3 or equivalent-concentration HCl. Figure 3.4 shows a typical separation of Na^+, NH_4^+, and K^+ ions. Ammonium is separated by both the separating resin and the suppressor resin. The eluents used for monovalent cation separations should be made from ultrapure acids. These chemicals have lower levels of impurities, such as transition metals, than reagent-grade chemicals. Impurities in the eluent can accumulate on the separating resin, thus reducing the available ion-exchange capacity. Eluent flow rates range from 2.3 to 4.5 mL/min. Use of HNO_3 rather than HCl as the eluent requires longer regeneration times because the NO_3^- is more strongly held, compared with the OH^- of the regenerant, than is Cl^-. However, system corrosion is *much* less with HNO_3 than with HCl eluents.

The same size separating column with packing identical to that in monovalent cation separations is used for divalent cation separations. The critical difference is the eluent. The $0.0025M$ m-phenylenediamine (termed m-PDA) mixed with a total of $0.0075M$ HCl, HNO_3, or $HClO_4$ has been effective in separating Mg^{2+}, Ca^{2+}, Sr^{2+}, and Ba^{2+} (Figure 3.5). The alkali metals are eluted early in the chromatogram as a single peak or a partly resolved double peak. The second peak in the doublet is ammonium

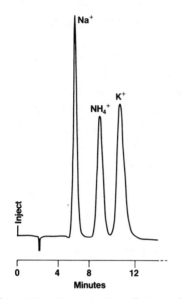

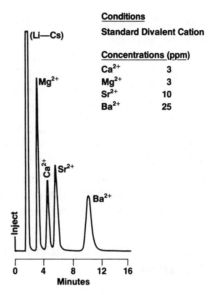

Figure 3.4. Determination of monovalent cations. Copyright by Dionex Corp. and reprinted by permission of copyright owner.

Figure 3.5. Separation of divalent cations. Copyright by Dionex Corp. and reprinted by permission of copyright owner.

ion. *m*-Phenylenediamine as the dihydrochloride salt or diamine should be obtained in the highest purity possible (no reagent-grade chemical is presently available). The eluent will undergo slow light and oxygen-induced degradation and should be carefully checked before each use. Fresh eluent should be prepared weekly. The eluent becomes dark or colored as it degrades; this also causes colors to appear in the separator and suppressor columns. Apparently contaminants and degradation products interact mainly with the resin backbone polymer because the discoloration does not affect column performance adversely.

If the *m*PDA/mineral acid eluent is used with a 6 × 250 mm cation separator column, separation of Na^+, NH_4^+, and K^+ will be significantly degraded. Daily washing of the separator column with $1M$ HNO_3 and water will sometimes delay the loss of separation of NH_4^+ and K^+.

The suppressor column used in both standard cation IC conditions is a 9 × 250 mm glass or a 9 × 100 mm plastic column. The typical elapsed time between regenerations is 4 to 6 h. Typically, the 30-min regeneration is followed by a 40-min water wash to reactivate the suppressor resin. If HNO_3 or *m*-PDA is the eluent, longer regeneration with the $0.5M$ NaOH is required to regain as much suppression capability as possible.

3.4 STANDARD ANION ION EXCLUSION (ANION IE) CONDITIONS

Because anion IE is a single-column system, instrument conditions are simple. The separating column is a 6 × 500 mm glass or 9 × 150 mm plastic column. Water is used as the eluent at a 2.3-to-3 mL/min flow rate. No precolumn or suppressor column is used. The separating column separates the strong acid anions (Cl^-, SO_4^{2-}, NO_3^-, etc.) in a single peak from acetate, formate, and carbonate ions (Figure 3.6). The weak acid peaks are typically sawtooth-shaped rather than nearly Gaussian. This system has been successful in coupled chromatography applications, for on-column sample pretreatment or actual ion separations of weak acids.

3.5 STANDARD ANION CHROMATOGRAPHY EXCLUSION CONDITIONS

Like anion IE, anion ICE separates ions with mechanisms that are not truly ion exchange. The anion ICE system is more complex than the anion IE system. Figure 3.7 shows a typical ICE separation.

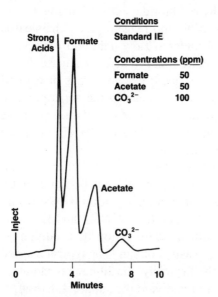

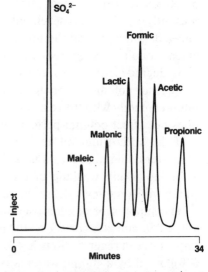

Figure 3.6. Ion exclusion separation. Copyright by Dionex Corp. and reprinted by permission of copyright owner.

Figure 3.7. Standard anion ICE determination. Copyright by Dionex Corp. and reprinted by permission of copyright owner.

The ionic species are fully protonated with the $0.0001M$ HCl eluent. This enhances separation by ICE. A 9×250 mm glass or a 9×150 mm or 9×300 mm plastic separator produces chromatograms typified by Figure 3.7. Eluent flow rate is kept very low (0.9 mL/min) because the separating columns have high pressure drops.

A disposable plastic silver-form suppressor column (6×150 mm) is used to convert the eluent to H_2O. The eluent is suppressed in a column filled with Ag^+ form resin which precipitates the Cl^- and exchanges the H^+ ions. A post-suppressor column filled with $\frac{1}{2}$ Ag^+ form/$\frac{1}{2}$ H^+ form resin enhances ion sensitivity. Because the precipitation of AgCl causes a continuing pressure increase as the suppressor column is used, the expended portion of the suppressor (gray-colored) is carefully cut off after each day of operation. When the column length becomes too short, the suppressor is replaced with a packed plastic refill. The post-suppressor usually does not need to be replaced for several months.

3.6   COUPLED IC CHROMATOGRAPHY CONDITIONS

Numerous combinations of different types of ion chromatographic column sets can be used in the coupled chromatography operating mode. Among these possibilities, the most promising right now are the following:

1. Anion IC/anion IC with two different eluents/columns.
2. Anion IE/anion IC.
3. Anion ICE/anion IC.
4. Anion IC/anion ICE.

Coupling two separation systems together usually increases the specificity of the ion analysis. Analysis times are longer with coupled chromatography. The initial system can also be used as an on-line pretreatment for the sample.

Several instrument features are required to operate in the coupled chromatography mode. First the two chromatographic systems should function independently. This usually requires two complete pumping systems. At least one detector is required, although use of dual detectors is easier. A valve to divert the regions of interest from the first system to the second is required. Also, at least one column is used to trap the ions of interest from the first system onto the second chromatographic system for subsequent analysis.

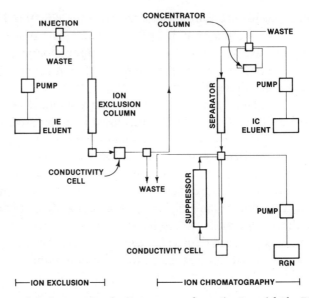

Figure 3.8. Coupled chromatography instrument schematic. Copyright by Dionex Corp. and reprinted by permission of copyright owner.

Figure 3.8 is a flow diagram for the coupling of an anion IE/anion IC system. The sample is injected only on the IE column. Because the strong acid anions are excluded from the IE separator resin, they elute in the void volume peak. These ions are detected by the first conductivity cell/meter. After detection they are diverted onto an anion-concentrating column located at the injection valve of the second system. After the IE void volume peak is trapped the divert valve is switched so that the other eluted peaks from the first system go to waste after detection rather than onto the trapping column. After the volume void peak is trapped on the concentrator column in the standard anion IC system, the 0.003M Na-HCO$_3$/0.0024M Na$_2$CO$_3$ eluent flow is changed so that it passes through the concentrating column. This injects the trapped peak for analysis under a different set of analytical conditions. Figure 3.9 is an example of coupled IE/IC analysis for a standard prepared in 4% NaOH.[1] The IE system neutralizes the NaOH and separates the strong acid anions from CO$_3^{2-}$ ion. These strong acids are subsequently analyzed by conventional anion IC for Cl$^-$, ClO$_3^-$, and SO$_4^{2-}$.

The anion ICE system has also been coupled to anion IC. The anion ICE system provides better separation and detection of organic acids than

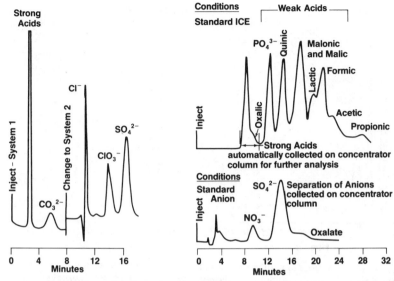

Figure 3.9. Separation of anions in 4% caustic. Reprinted with permission of *Ann Arbor Science.*

Figure 3.10. Strong and weak acids in coffee extract. Copyright by Dionex Corp. and reprinted by permission of copyright owner.

the IE system. Figure 3.10 is an example of anion ICE/anion IC analysis of coffee. Once again the sample is injected through the ICE system first. Sulfate and oxalate ions, as well as a variety of organic acids, are observed in the ICE chromatogram. The region that contains the sulfate and oxalate peaks in the ICE system is trapped for standard anion IC analysis. These two ions elute quite closely together under anion ICE conditions. By subsequent IC analysis they are well separated, as shown in the lower portion of the chromatogram.

3.7 NONSTANDARD OPERATING CONDITIONS

The ionic composition of the sample is the main determining factor in choosing instrument conditions for ion separations. If the sample is a complex mixture of ions with widely varying affinities for the resin, several sample injections may be necessary. They may require the use of different eluting ions, eluent strengths, or even different separating columns or column lengths.

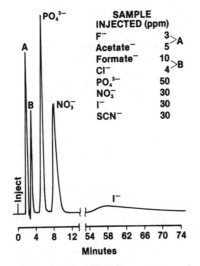

Figure 3.11. Multianion mixture eluted with standard anion conditions. Copyright by Dionex Corp. and reprinted by permission of copyright owner.

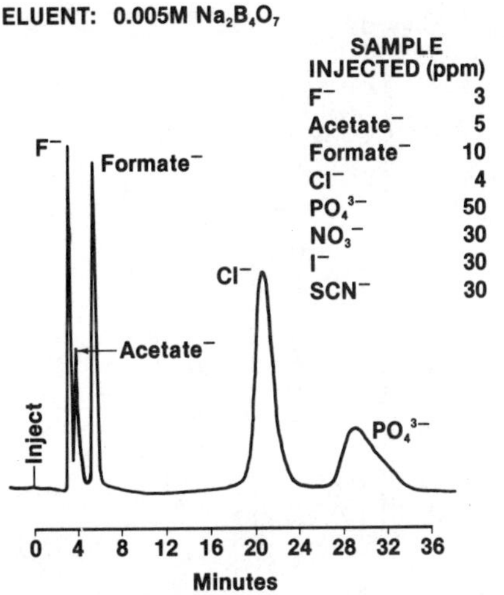

Figure 3.12. Multianion mixture eluted with borate eluent. Copyright by Dionex Corp. and reprinted by permission of copyright owner.

A good example of a complex sample with a wide range of ion retentions is the anion separation of a sample that contains fluoride, acetate, formate, chloride, phosphate, nitrate, iodide, and thiocyanate ions. Figure 3.11 shows this sample eluted under standard anion IC conditions. Only the PO_4^{3-} and NO_3^- are easily quantitated in this run. The acetate and fluoride co-elute, as do the chloride and formate. Iodide is too late-eluting and broad to be accurately measured. Thiocyanate does not elute as a peak under these instrument conditions. Thus only two ions in this sample can be quantitatively determined under standard conditions. The instrument conditions must be changed to obtain improved separations for the other ions.

The weakly held ionic species can be separated by a different eluting ion. Figure 3.12 shows the same sample run on the same separator with $0.005M$ $Na_2B_4O_7$ to separate the weakly held species. The separator column reequilibration is aided by injecting high concentrations of $B_4O_7^{2-}$ (a low affinity ion) to displace the HCO_3^- and CO_3^{2-} ions as the counterions at the active site of the resin. At the same time the suppressor column should be regenerated. After the separator is in the $B_4O_7^{2-}$ form (made evident by a stable baseline) the sample is reinjected. With the eluent used the fluoride, acetate, and formate peaks nearly coelute, and chloride

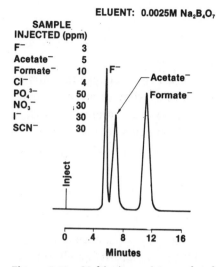

Figure 3.13. Multianion mixture eluted with weaker borate eluent. Copyright by Dionex Corp. and reprinted by permission of copyright owner.

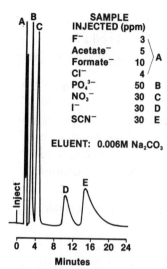

Figure 3.14. Multianion mixture eluted with strong carbonate eluent. Copyright by Dionex Corp. and reprinted by permission of copyright owner.

and phosphate ions can be measured with ease. The NO_3^-, I^-, and SCN^- ions all remain on the separating column, thus reducing the effective separator capacity for subsequent runs. Figure 3.13 shows an improved separation of acetate and fluoride when the eluent strength is halved. With $0.0025M$ $Na_2B_4O_7$ eluent the chloride and phosphate peaks do not elute in a reasonable time.

Quantitative analysis of the strongly retained ions I^- and SCN^- requires a different column and a higher concentration of CO_3^{2-} eluent. Figure 3.14 shows the separation of I^- and SCN^- with $0.006M$ Na_2CO_3 and a 3×250 mm glass L-20 separator column. This column was used in these separations to reduce peak tailing in the I^- and SCN^- peaks. Cyanophenol has been used as an eluent additive to eliminate the need for the L-20 column. Other columns have been introduced that perform the same task. The $0.006M$ CO_3^{2-} elutes the weakly retained ions in several poorly resolved peaks early in the chromatogram. Phosphate and nitrate along with iodide and thiocyanate, can be determined quantitatively.

As the preceding example shows, instrument operating parameters can be changed in a variety of ways when standard conditions are insufficient. Three main aspects of an IC separation can be optimized:

1. Resolution of detected ions.
2. Speed of analysis.
3. Sensitivity or detectability.

Each can be changed in a variety of ways; however, a change to optimize one parameter usually affects the other two. A variety of means is available in ion chromatographic separations to alter each parameter, most of which simply involve minor instrument adjustments, eluent or column changes, or sample prechemistry. Some involve more complex plumbing or eluent modifications.

3.8 INCREASING RESOLUTION

A frequent problem in any type of chromatography is the need to increase resolution between two or more peaks. Once standard IC conditions are determined to be inadequate to resolve sample ions a variety of pathways have been developed to improve ion separations. In order of increasing complexity and time consumption these alternatives are (a) sample dilution or pretreatment; (b) use of different lengths or types of separator columns; (c) use of different eluent strengths or eluents; and (d) use of different ion-separating mechanisms.

One of the easiest methods of increasing resolution is to dilute or pretreat the sample. A good illustration is the analysis of brines for sulfate and nitrite. Figure 3.15 shows direct injection of a 10 ppm SO_4^{2-} standard in 10% NaCl. The separator column is overloaded, as revealed by the broad peak width and excessive tailing of the normally sharp chloride peak. The column is still eluting chloride ion after ½ h. No other sample should be injected until a stable baseline is obtained. When the same sample is diluted 10-fold and reinjected the 1-ppm sulfate peak is easily detectable (Figure 3.16).[2] Detection limit for SO_4^{2-} in brine is about 100 ppb SO_4^{2-} in 1% NaCl. The 10-fold dilution produced a good ion separation of Cl^- and trace SO_4^{2-} without sacrificing sensitivity and without overloading the column set. Practical experience indicates that the maximum recommended ion concentration for any anion is about 500 μg (5000 ppm with a 0.1-mL injection). When this amount is exceeded, the column set may well become overloaded with sample ions.

Dilution is an effective method of obtaining detectable sulfate peaks in brines because of the large separation factor between chloride and sulfate. It is not effective in obtaining reasonable nitrite determinations

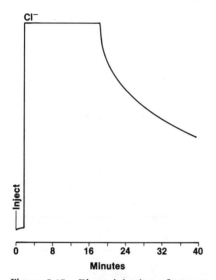

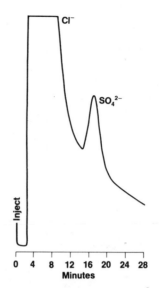

Figure 3.15. Direct injection of 10 ppm SO_4^{2-} in 10% NaCl brine. Copyright by Dionex Corp. and reprinted by permission of copyright owner.

Figure 3.16. Analysis of 1 ppm SO_4^{2-} in 1% NaCl. Copyright by Dionex Corp. and reprinted by permission of copyright owner.

in brine. The nitrite and chloride have similar affinities for the resin. Even after dilution the excessive width of the chloride peak still obscures the nitrite peak.

Another method of nitrite determination in the presence of high chloride concentration is the chemical pretreatment of the sample. Several pretreatments to remove excess chloride ion have been developed; for example, the sample can be injected through a silver-form precolumn to precipitate the chloride ion. Silver oxide can also be added to the sample before the injection; the nitrite ion is unaffected by the silver and chloride ion is precipitated. The precipitated AgCl is removed by filtration before the sample is injected into the IC. Figure 3.17 illustrates a sample after silver precolumn treatment. Note that sulfate and phosphate as well as nitrite can be separated after the pretreatment but that the other halides are removed.

Care must be taken in any pretreatment procedure to avoid introducing ionic contaminants into the sample, nor must pretreatment remove any of the ions of interest. When attempting a chemical treatment of the sample it is essential to run a blank that closely resembles the sample and to check recovery of the ions of interest in a series of spiked samples.

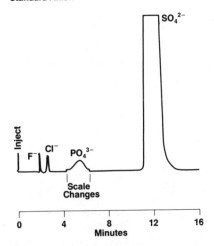

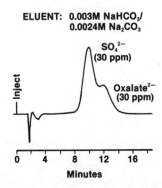

Figure 3.17. Anions in ocean water after chloride removal. Copyright by Dionex Corp. and reprinted by permission of copyright owner.

Figure 3.18. Elution of sulfate and oxalate under standard anion IC conditions. Copyright by Dionex Corp. and reprinted by permission of copyright owner.

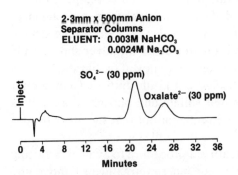

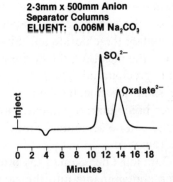

Figure 3.19. Use of longer anion separator column to improve sulfate–oxalate resolution. Copyright by Dionex Corp. and reprinted by permission of copyright owner.

Figure 3.20. Optimized sulfate–oxalate separation. Copyright by Dionex Corp. and reprinted by permission of copyright owner.

A second method of achieving better resolution is to increase the separating column length. This will result in improved separations, although analysis times are longer and system-operating pressure is higher. Figure 3.18 shows the partial separation of sulfate and oxalate ions run under standard operating conditions. A single 3 × 500 mm glass anion separator column produces partial separation with a 15-min run time. Use of two 3 × 500 mm glass anion separator columns (Figure 3.19) gives baseline resolution for these ions. However, system-operating pressure is doubled and run time is nearly twice as long. A stronger eluent ($0.006M$ Na_2CO_3) used with these 2 columns elutes the well resolved oxalate and sulfate peaks in a shorter time while maintaining excellent resolution (Figure 3.20).

The size of the suppressor column can also affect ion separation and reproducibility. Nitrite ion interacts with the anion suppressor resin, apparently by adsorption; the water dip is also a problem in nitrite analysis, especially when a freshly regenerated suppressor column is used. Repetitively injected nitrite standards show significant increases in peak height.[3] The half-width of the nitrite peak slightly decreases as the suppressor column is expended. In one solution to the nitrite analysis problem two 3 × 150 mm anion suppressor columns are used in parallel.[4] These columns are alternated for each sample. Every time a suppressor column is changed the off-line column is run through a short regeneration cycle,which results in the nitrite ion continually interacting with a fully regenerated bed volume of suppressor resin. Naturally an automated IC is desirable if this method is to be used routinely. Changing the type of separator column will sometimes enhance separation of poorly resolved ions. Several different separators are available for suppressor-type anion separations. These columns vary in the functional group on the latex particle, in the particle size of the latex bead, or in the cross-linking of the latex bead. Specific information on these differences is not readily available. Two good examples are the use of the L-20 separator column (covered earlier) in the separation of I^-–SCN^- and the use of the SA-2 separating columns (from Dionex) to enhance Cl^-–NO_2^- separation or to alter Br^-–NO_3^- separation. Few variations in silica ion-exchange columns used in nonsuppressed IC applications are available.

One effective method of increasing peak separation involves changing eluents. Chromatographic separations are based on competition between eluent ions and sample ions for the active capacity of the resin. To compete effectively sample ions and eluent ions should have similar affinities for the resin within a factor of about 2 to 5. The standard condition eluents are generally best suited to separate moderately retained ions such as sulfate and potassium. When these eluents do not yield acceptable separations for the various sample ions, other conditions can be used to improve separation. As a general rule fairly dramatic changes in eluent concentration or eluting ion type are required to alter ion separations. Eluent selection principles are illustrated by the following examples.

Ions that elute before Cl^- under standard anion conditions are considered to be weakly retained. These ions are generally monovalent inorganic anions, monofunctional carboxylic acids and other organic species, and many weakly ionized species such as HCO_3^-, CN^-, and S^{2-}. Separation of these weakly held species can be increased by using one of several weak eluting ions. Especially useful for separating them are bicarbonate ion (HCO_3^-), hydroxide ion (OH^-), and tetraborate ion ($B_4O_7^{2-}$). Figure 3.21 shows the elution of fluoride and chloride ions under standard anion IC conditions. Figure 3.22 illustrates the effect of a change to $0.0015M$ $NaHCO_3$ on this separation. Use of a hydroxide eluent will result in a similar separation because the hydroxide ion is equal in eluting strength to the HCO_3^- ion. With either eluent fluoride is still eluted at nearly the same retention time, but the higher affinity of the chloride in relation to HCO_3^- or OH^- ion causes this sample ion to elute at about 12 min. If $0.005M$ $Na_2B_4O_7$ is used as the eluent, not only can a similar separation between fluoride and chloride be obtained but formate and acetate ions will also elute between the two halides (Figure 3.23). As this eluent strength is decreased, acetate and formate are better separated, but chloride elutes after 30 min. Care must be taken when preparing NaOH and $Na_2B_4O_7$ eluents to insure that dissolved CO_2 gas (which forms CO_3^{2-} and HCO_3^- ions in solution) is removed from the water in which the eluents are prepared. If the gas is not removed, the much higher eluting strength of the CO_3^{2-} ion causes irreproducible results from batch to batch of eluent.

Moderately retained anions are usually best separated under standard conditions or with other moderate strength eluents. Figure 3.18 shows the separation of oxalate and sulfate under standard conditions. When the eluent strength is halved (Figure 3.24) no increase in resolution is observed but the run time is twice as long. This separation is optimized by using two 3 × 500 mm anion separator columns with $0.006M$ Na_2CO_3 (Figure 3.20). The higher strength Na_2CO_3 will expend the suppressor at a slightly faster rate than the standard eluent.

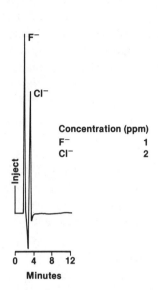

Figure 3.21. Elution of F^- Cl^- under standard anion IC conditions. Copyright by Dionex Corp. and reprinted by permission of copyright owner.

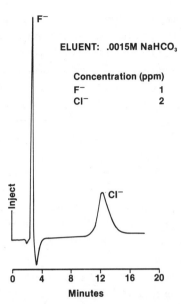

Figure 3.22. Increased F^- Cl^- separation by use of a weaker eluent. Copyright by Dionex Corp. and reprinted by permission of copyright owner.

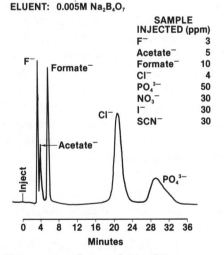

Figure 3.23. Elution of F^-, Cl^-, acetate, and formate. Copyright by Dionex Corp. and reprinted by permission of copyright owner.

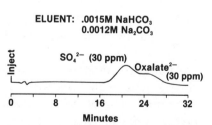

Figure 3.24. Effect of decreased eluent strength on oxalate–sulfate separation. Copyright by Dionex Corp. and reprinted by permission of copyright owner.

45

One interesting example of an eluent change that drastically alters an elution profile of moderately retained ions makes use of mixed OH^-/CO_3^{2-} eluents to elute phosphate ion after sulfate. Figure 3.25 shows the separation of ions in boiler blow-down waters with the combined CO_3^{2-}/OH^- eluents and two 3 × 500 mm glass columns.[5] The higher pH of these eluents changes the form of the orthophosphate from the monovalent $H_2PO_4^-$ ion eluted in standard conditions to the divalent HPO_4^{2-}. The divalent form has a stronger affinity for the resin and thus elutes later.

Di- or trivalent inorganic ions, polyfunctional carboxylic acids, and metal oxide anions are generally more strongly held than SO_4^{2-}. Some monovalent species, however, such as I^-, SCN^-, and ClO_4^-, are also strongly retained. These ions have large hydrated ionic radii which may prevent eluent ions from approaching the exchange sites where these ions are exchanged.

Moderate-strength carbonate eluents are generally ineffective in separating high affinity species. The iodide and thiocyanate can be well separated by using higher strengths of CO_3^{2-} eluting ion if p-cyanophenol is used as an adsorption modifier. Figure 3.26 is an example of these two ions completely separated in 1% NaCl. Perchlorate ion is much more strongly retained than I^- or SCN^-. The use of $0.005M$ NaI eluent with silver suppression has resulted in good separation of ClO_4^-, even in 1% NaCl (Figure 3.27). The Cl^- peak is small because almost all the chloride ion is precipitated in the silver suppressor. Another eluent for these strongly retained ions is citrate ion. Pyrophosphate and tripolyphosphate ions have been eluted with $0.01M$ sodium citrate (Figure 3.28). This eluent has a high conductivity background even when a suppressor column is used and care should be taken to minimize impurities in the chemicals chosen to prepare the eluent. Problems with column materials degrading with prolonged use of this eluent have been reported.[6] This degradation may originate in the samples or in bacterial growth in the eluent or columns.

Another method of increasing resolution between two peaks that closely elute with standard conditions involves a different separating mechanism. Use of the ICE separation method will also alter the elution profile of anionic species significantly. The separations obtained on this column set are based on non-ion-exchange mechanisms and elution patterns considerably different than those of standard anion IC are observed.

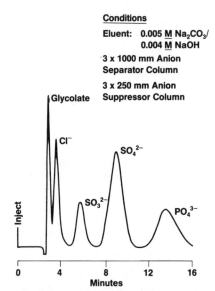

Conditions

Eluent: 0.005 $\underline{M}$ Na$_2$CO$_3$/
0.004 $\underline{M}$ NaOH

3 x 1000 mm Anion
Separator Column

3 x 250 mm Anion
Suppressor Column

Glycolate

Cl$^-$

SO$_4^{2-}$

SO$_3^{2-}$

PO$_4^{3-}$

Inject

0   4   8   12   16
Minutes

Figure 3.25. Anions in boiler blow-down water. Reprinted from *Analytical Chemistry* with permission.[5] Copyright 1977 American Chemical Society.

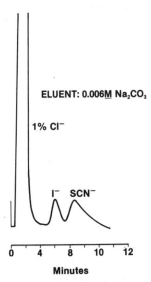

ELUENT: 0.006$\underline{M}$ Na$_2$CO$_3$

1% Cl$^-$

I$^-$ SCN$^-$

0   4   8   10   12
Minutes

Figure 3.26. Iodide-thiocyanate separation in 1% NaCl. Copyright by Dionex Corp.; reprinted by permission of copyright owner.

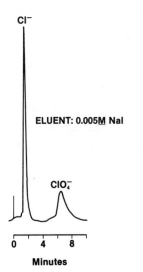

Cl$^-$

ELUENT: 0.005$\underline{M}$ NaI

ClO$_4^-$

0   4   8
Minutes

Figure 3.27. Perchlorate in brine. Copyright by Dionex Corp. and reprinted by permission of copyright owner.

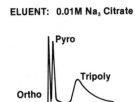

ELUENT: 0.01M Na$_3$ Citrate

Pyro

Tripoly

Ortho

Figure 3.28. Separation of ortho-, pyro-, and tripolyphosphates.

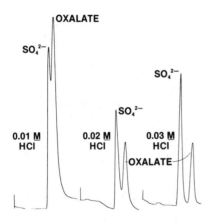

Figure 3.29.   Effect of pH on oxalate-sulfate separation by ICE. Copyright by Dionex Corp. and reprinted by permission of copyright owner.

It appears that ICE elution profiles depend primarily on the pK_a of the eluted ions; species with higher pK_a elute later. Adjustment of pH will affect separations significantly. Figure 3.29 shows the separation of SO_4^{2-} and oxalate by ICE with increasing HCl strengths. As the pH goes down the separation improves. However, the suppressor column is expended much faster as the HCl strength increases. Other changes, such as lowering flow rate and lengthening the separator column, are not usually effective in the ICE system because of column and instrument limitations.

3.9   DECREASING ANALYSIS TIMES

The second aspect of IC analysis that can be optimized for a particular ion separation is the amount of time it takes to determine the species of interest. Shorter analysis times are naturally better if adequate results can be obtained, but shortening the analysis time is often in direct conflict with resolution requirements. A number of operating parameters can be changed to alter the analysis time for a sample:

1.   Separator column capacity.
2.   Eluent flow rate.
3.   Eluent strength.
4.   Use of organic modifiers in the eluent.
5.   Use of gradients.
6.   Use of separator backflushing methods.

One of the easiest approaches to shortening analysis time is to lower the separating capacity. This is best achieved by using separating columns of shorter length. If nitrate and sulfate ions are to be determined in a water sample, a 3 × 500 mm glass anion-separating column will give excellent separation of these ions with an 18-min analysis time. A 3 × 250 mm glass anion separating column (one-half the capacity of a 3 × 500 mm column) will lower the analysis time to about 9 min, but the nitrate and sulfate will not be resolved to baseline. The separation is still adequate to obtain quantitative values for both ions. Use of a different eluent will give baseline resolution for these ions with an 8-min run time. Use of too short a column will not provide adequate resolution, even if the eluent is changed. A 3 × 150 mm column will separate the sulfate from the nitrate ion only slightly and is not adequate for routine use.

Eluent flow rate can be increased to reduce sample analysis time. The main restriction on eluent flow rate is maximum system pressure limits. With Tefzel tubing the present system practical operating limit is about 2,000 psig. If the IC is operated at higher pressures, a liquid line may rupture, one of the gripper fittings may pull loose (causing a liquid leak), or more likely a column or valve will leak. As the instrument flow rate is increased, resolution between adjacent peaks decreases along with analysis time. Non-ion-exchange separations are especially affected by changing eluent flow rates. A severe decrease in resolution between Br^- and NO_3^- (whose separation is not entirely due to ion exchange) is observed at higher flow rates, whereas $NO_3^- - SO_4^{2-}$ are still well resolved even at high flow rates.

Another method of altering sample analysis time is to change the eluent strength. Increasing eluent strength usually has the same effect as shortening the separator column or raising the flow rate—proportionally shorter run times at the expense of overall resolution. Use of stronger eluting ions (polyvalent ions such as citrate and m-phenylenediamine) will also cause ions to elute faster but with loss of resolution for weakly or moderately retained ions. Use of weaker eluting ions may result in better separation of weakly retained ions, but some species may elute with long retention times or not at all.

Organic additives which reduce peak tailing have been added to the CO_3^{2-} eluent for improved I^- and SCN^- separations in 1% NaCl. Because the $I^- - SCN^-$ separation is due in part to adsorption, the use of p-cyanophenol for adsorption site blocking reduces peak tailing and thus enhances separation. This modifier can also adjust the elution characteristics of Br⁻–NO₃⁻ on an anion separator column. However, resolution between these ions is usually lost when this additive is used. A number of other possible modifiers are being investigated to obtain improved separations for certain ions.

The last two methods of decreasing analysis time are considerably more complex and have been used only for special applications. Gradient elution is usually accomplished by steps rather than by a continuous gradient. Gradient elution has one major problem: the detector follows the increase/decrease in eluent strength. Constant conductivity gradients have found no common use because the chromatogram baseline is quite noisy and has considerable drift. Finally, backflushing or column-switching methods have been used to elute strongly retained ions after the weakly and moderately held ions have been eluted.

Low, medium, and high affinity anions can be separated in a single injection by gradient elution. This requires a three-step gradient with varying concentrations of Na_2CO_3 as the eluent (Figure 3.30).[7] High eluent flow rates minimize eluent changeover times. The first eluent is an extremely weak CO_3^{2-} eluent (0.00015M) to separate F^-, CHO_2^-, $CH_3CO_2^-$, Cl^-, and NO_2^-. Before the chloride peak elutes the second eluent (0.0018M Na_2CO_3) is introduced. Because of the dead volume in the transfer lines, this eluent does not pass through the separator and

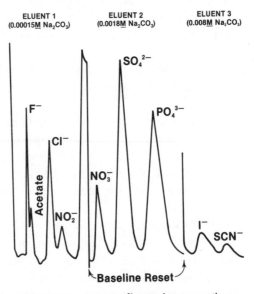

Figure 3.30. Step gradient anion separation.

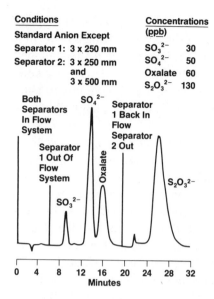

Conditions

Standard Anion Except

Separator 1: 3 x 250 mm

Separator 2: 3 x 250 mm
and
3 x 500 mm

Concentrations (ppb)

SO_3^{2-} 30
SO_4^{2-} 50
Oxalate 60
$S_2O_3^{2-}$ 130

Both Separators In Flow System

Separator 1 Out Of Flow System

SO_4^{2-}

Separator 1 Back In Flow

Separator 2 Out

Oxalate

SO_3^{2-}

$S_2O_3^{2-}$

0 4 8 12 16 20 24 28 32
Minutes

Figure 3.31. Single-injection analysis of SO_3^{2-}, SO_4^{2-}, $S_2O_3^{2-}$, and oxalate. Copyright by Dionex Corp. and reprinted by permission of copyright owner.

suppressor column until the NO_2^- peak has eluted. A large jump in conductivity background requires an increase in the electronic offset to get an electrically zero baseline. The second eluent elutes NO_3^-, PO_4^{3-}, and SO_4^{2-} ions. No resolution between Br^- and NO_3^- is seen with this eluent. The third eluent is introduced after the SO_4^{2-} peak maximum. The I^- and SCN^- peaks show quite a bit of tailing with the $0.008M$ Na_2CO_3 third-step eluent. Use of p-cyanophenol in the third eluent should reduce peak tailing.

Another gradient system used for this same type of separation involves switching of columns to eliminate duplicate sample injections while using isocratic elution. The plumbing on the ion chromatograph must be changed to allow switching of the separator columns. Figure 3.31 is a chromatogram of the separation of SO_3^{2-}, SO_4^{2-}, oxalate^{2-}, and $S_2O_3^{2-}$. The sample is first injected with both columns in-line. The first separating column (a 3×250 mm glass column or equivalent) holds up the strongly retained $S_2O_3^{2-}$ ion. The moderately retained species (SO_3^{2-}, SO_4^{2-}, and oxalate^{2-}) are fairly rapidly eluted from the first separator with normal eluent (about 6 min). The first column switch is then made. The 3×250 mm first separator is switched out of the eluent stream while the 3×750 mm second separator column stays in-line to separate the moderately held ions. The $S_2O_3^{2-}$ is retained on the first separator column for later elution. After the three moderately retained ions are eluted from the second separator and are detected, the second column switch is made. The longer second separator column is switched out of the eluent stream and the first column (holding the $S_2O_3^{2-}$) is placed back in-line. Eluent is then pumped through the first 3×250 mm column to elute the $S_2O_3^{2-}$. Great care must be taken to allow enough eluent to flow through both columns to elute the first three ions completely while keeping the $S_2O_3^{2-}$ quantitatively on the first separator column.

Sometimes eluent and column gradients must be combined to achieve good separations. Monovalent and divalent cation separations require two separator columns and entirely different eluents. In a tentative method column and eluent switching separate the monovalent and divalent cations in a single injection. This method has only been reported[8]; it is not commonly used. The two eluents needed are those for monovalent (Na^+, NH_4^+, K^+) and divalent (Mg^{2+}, Ca^{2+}) cation separations (see earlier sections on standard condition cation IC separations). The strongly retained ions (Mg^{2+} and Ca^{2+}) are held up on the first separator while the moderately retained ions are passed through both separators. The first 3×500 mm cation separator column is switched out of line shortly after injection (still holding the Mg^{2+} and Ca^{2+}). The monovalent cation analysis is then completed for Na^+, NH_4^+, and K^+ on the second separator, a 6×250 mm cation separator column. The eluent is $0.005M$ HNO_3, which is changed to protonated *m*-PDA at the same time the second separator column is switched out of the eluent stream. Concurrently, the first column is placed back in-line. This elaborate switching arrangement is required to prevent phenylenediamine from getting onto the monovalent cation separator column. If this should occur, it could result in NH_4^+–K^+ separation loss. No detector offset change is needed because the background conductivity of both eluents is close to zero. A problem with sodium retention by the divalent separator has been observed and must still be solved. Adequate recovery studies for Ca^{2+} and Mg^{2+} also have not been completed.

Figure 3.32 shows a chromatogram from an IC modified for sequential ppm and ppb level determinations. The normal instrument configuration is altered to isolate both injection systems. This isolation is important to minimize cross-contamination.

A blackflushing-instrument modification has been used to elute strongly retained species (i.e., CrO_4^{2-}) from the top portion of the separator column after separation and detection of moderately held anions such as $H_2PO_4^-$. The liquid plumbing must be modified to allow reverse

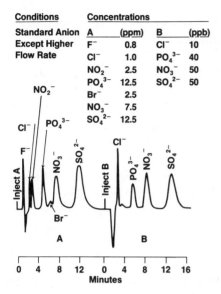

Conditions	Concentrations			
Standard Anion	A	(ppm)	B	(ppb)
Except Higher	F^-	0.8	Cl^-	10
Flow Rate	Cl^-	1.0	PO_4^{3-}	40
	NO_2^-	2.5	NO_3^-	50
	PO_4^{3-}	12.5	SO_4^{2-}	50
	Br^-	2.5		
	NO_3^-	7.5		
	SO_4^{2-}	12.5		

Figure 3.32. Trace and ultratrace anions with modified IC. Copyright by Dionex Corp. and reprinted by permission of copyright owner.

flow of the eluent in the separator column.[9] Very little baseline disturbance is noted when the column is backflushed.

These advanced methods for ion separation have not yet been widely incorporated into IC methods. Most work has been done to demonstrate the feasibility of using the method and not for routine sample analysis. New columns and eluents are being developed to handle these problems. The need for careful attention to electronic offset and eluent changes almost mandates an automatic ion chromatograph for routine work if these advanced methods are used.

3.10   CHANGING DETECTABILITY OR SENSITIVITY

Response for each ion depends on its dissociation in the suppressor column effluent and on the ability of the various species in the conductivity cell to transmit electrical current. Sensitivity is defined as the detector response per unit ion concentration. For totally dissociated species, such as HCl and H_2SO_4, strong signals are obtained. For weaker species like H_2CO_3 little response from the conductivity detector is observed, even on a water background. There is a substantial difference in sensitivity if organic solvents are used in the eluent. Most solvents inhibit ionization and lower detector output. Detectability, or signal-to-noise ratio at low ion concentration levels, is dependent on the sensitivity (i.e, the response per unit concentration) of a particular ion.

Ion detectability can be increased or decreased in a variety of simple ways. One of the easiest is to use the lowest numerical scale settings on the conductivity meter. Recorded peaks from these high instrument sensitivity settings are more easily and accurately measured. All that is necessary for this to work is that the species must already be detectable above instrument background noise which must not be excessive. The overall system noise quickly becomes the limiting factor in low-level ion determinations. Several steps can be taken to minimize background noise, among which are the elimination of air bubbles and minor liquid leaks in the valves. Use of a good pulse damper (such as a long length of large-diameter Teflon tubing connected to the gauge or before the injection valve) or a pulseless pump will minimize the noise from the pump. Both steps will lower the effective noise level and enhance detectability.

Another simple method of improving ion detectability is to use a larger injection volume. The practical upper limit for direct injection in anion systems is between 500 and 1000 µl. The "water dip" becomes so wide with large volume injections that early eluting species are often obscured in the dip. Larger volumes can be used for cation systems. A good example in which this has worked well is the determination of 10 ppb Na^+ in 2 to 3 ppm of NH_4OH. If a concentrator column is used, the large amount of NH_4^+ lowers the column affinity for Na^+ ions. Sodium recovery is not quantitative. The solution to this problem is a large injection volume. A 10-mL sample volume is injected through the precolumn and separator. Figure 3.33 shows the separation of 10 ppb Na^+ in a 350-fold excess of NH_4^+ ion.[10]

A more complex method for improving detectability requires a concentrator column. These small columns were designed to work with ultratrace levels of ions in clean samples (Figure 3.34),[11] but their use has been extended to much more complex samples.[12] The concentrator is used to strip ions out of large sample volumes that are pumped through the column. Typically, sample volumes range from 10 to 50 mL. Better reproducibility is obtained when the samples are loaded from a pump rather than by hand and when the sample loading and eluent flow are in the same direction. Loading the concentrator is independent of the loading flow rate. Samples have been loaded at remote locations and then transported to the instrument location, but reproducibility of the subsequent IC analyses is poorer than when the samples are loaded at the instrument

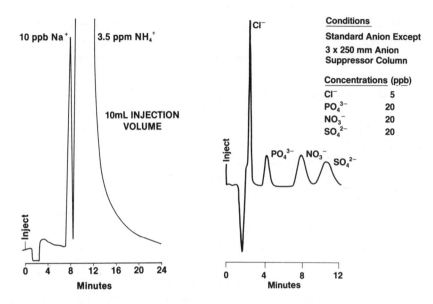

Figure 3.33.   Large-volume cation injection for trace Na⁺ analysis. Copyright by Dionex Corp. and reprinted by permission of copyright owner.

Figure 3.34.   Ultratrace levels of anions in power plants. Copyright by Dionex Corp. and reprinted by permission of copyright owner.

site. Ultratrace samples are best stored in specially conditioned polystyrene containers for less than eight days.[13]

Several points must be considered when a concentrator column is used. First, its loading capacity is limited. The practical total ion-exchange capacity of the resin in the column is only 50 to 60% of the theoretical capacity. Several causes have been postulated. Areas in which the sample ions can flow rapidly past the resin beads (i.e., channels, areas of poorly packed resin, areas near the walls) have poor exchange efficiency. This lack of exchange capacity is also observed when sample flow is restricted (where the beads touch each other, where the resin is so tightly packed that the result is bead distortion or collapse, and where resin clumping acts as a flow barrier). Both factors work to limit sample-loading capacities on concentrator columns. Uneven application of the sample onto the resin bed may also contribute to the loading limitation.

Other ions in the sample must be carefully considered for their impact on elution profiles. Potential problems exist from dissolved gases such as CO_2 and O_2, from organic species such as humic and fulvic acids, from fouling or very strongly retained inorganic species (silicates, polyphosphates, etc.), and from alcohols and other organic solvents that may affect the concentrator resin bed adversely. Each can cause a variety of chemical problems that can make concentrator column results appear to be non-reproducible.

A poignant example of the need to consider the entire sample when using concentrator columns is the ultratrace analysis of chloride and sulfate in samples from power plant turbine-steam condensates. The initial feasibility of concentrator columns was determined with standards prepared from 1 to 10 ppb each of chloride and sulfate ions in water. These samples were loaded onto a 3×50 mm concentrator column, and 10 to 20 mL loadings produced peaks with sufficient height for easy measurement, even at these ultratrace levels. The presence of ammonia in the samples from the steam-driven turbines was known but not considered. Samples were run with constant volume loadings (per manufacturer's instructions) with no difficulties, but, when the loading volume was increased to improve detection, the plot of response versus loading volume was linear for sulfate; however, the same plot for chloride ion fell off when the loading volume was increased. After careful investigation it was discovered that the presence of NH_3 was the cause of the problem. Ammonia is added to the turbine waters to elevate pH, thus minimizing corrosion problems; it forms NH_4OH when mixed with H_2O. This naturally dissociates to NH_4^+ and OH^-. The hydroxide ion has an affinity for the concentrator resin similar to chloride ion, but both have a much lower affinity than SO_4^{2-}. The OH^- ions act as an eluent for the much lower concentration Cl^- ions. The total chloride loading was thus limited by the OH^- that eluted the chloride as the sample was loaded. The sulfate ion, with its much higher affinity for the resin, was not affected by the OH^- ion. The solution to this apparent problem is extremely simple; just raise the capacity of the concentrator column. This can be done with a 3×150 mm glass column or a 4×50 mm plastic column. With the higher capacity the Cl^- is quantitatively retained, even with high OH^- concentrations, a simple solution once the problem was recognized.

Several uses of concentrating columns for high-concentration samples have been developed. Chloride impurities in acetic acid have been analyzed by using concentrator columns.[14] Calcium and magnesium in brines have also been determined by IC and concentrator columns. These ions have been directly analyzed in high-concentration brines (greater than 10% NaCl) by slow, direct injection of 1 mL of sample through a $3 \times$ 150 mm concentrating column following by a 1-mL deionized water wash to remove the excess Na^+. The large Na^+ peak is followed by much smaller Mg^{2+} and Ca^{2+} peaks (Figure 3.35).[15]

A different detectability problem occurs when samples that are too concentrated are injected into an IC column set. An overloaded separating column can result in unusual peak shapes or total lack of component resolution. Fortunately the inherent ruggedness of the IC separator resins and their ability to handle overloading without damage usually means that total column destruction is not a problem. Still, some loss of resolution between poorly resolved peaks may occur or a long re-equilibration of the column in which there is no sample throughput may be required. There may also be some changes in separation capability for ions that previously were well separated.

The easiest solution is to dilute the sample. It is often best to dilute in eluent if dilution of 100-fold or more is required. This will balance the sample and eluting ions. Any adverse reactions between sample and eluent ions, such as precipitation or oxidation, take place outside the columns. This procedure is practical only with manual handling of the sample because adequate automatic diluters have not yet been developed. An alternative to sample dilution is to use smaller injection loops. Volumes as low as 7 to 10 μL have been made from small-diameter Teflon tubing. These loops tend to plug with particulates, making it necessary to filter the sample prior to loop loading. Instrument reproducibility seems

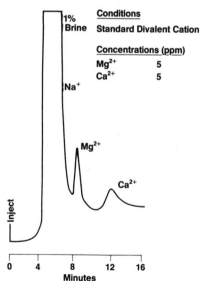

Figure 3.35. Divalent cations in 1% NaCl. Copyright by Dionex Corp. and reprinted by permission of copyright owner.

to be somewhat poorer with these small loops, possibly because of incomplete flushing of the loop or smearing of the sample ions on the injection valve slider. High viscosity samples and cross contamination in the injection valve may also be responsible for the losses in method reproducibility.

Several approaches have been used to improve instrument sensitivity (response per unit ion concentration). These include decreasing the distance between electrodes in the conductivity cell and using detection enhancing eluents, post-suppressor columns, and different detection methods in conjunction with the normal ion chromatography conductivity detector.

A reduction in the distance between the cell electrodes lowers the volume of the cell, causes an increase in the response for a given ion concentration in the cell, and increases the flow velocity through the cell. Although the instrument will be more sensitive for ions, the working range for reasonably linear calibration plots will be much more limited. The cell will also be more likely to plug with particulates and will not pass bubbles easily.

At least three auxiliary methods in IC, based on eluent, column, or detector changes, have been developed to detect ions with pK between 7 and 10. One involves modifying the eluent to produce detectable species. A good example is the ICE analysis of borates that uses $0.1M$ mannitol with $0.001M$ HCl. In another method a post-suppressor column changes the ionic form of the detected ions. Chloride-form post columns have been

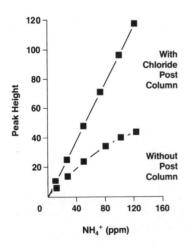

Figure 3.36. Linearity of NH_4^+ with chloride post column and without post column. Reprinted from *Analytical Chemistry* with permission.[16] Copyright 1977 American Chemical Society.

used successfully to improve the linearity of the ammonium ion calibration curve.[16] A comparison is made in Figure 3.36, but a 50% loss in sensitivity is apparent for ammonium.[17] Attempts have been made to use a sodium-form post column for cyanide analysis, but severe reproducibility problems were observed.[18] A third method to improve sensitivity for weak species use other detectors, one of which uses electrical resistivity (negative conductivity peaks).[19,20] Here the eluent is changed to provide a high conductance background. Electrical compensation for the background is applied at the recorder. This allows the use of high sensitivity scales on the meter. Tetraborate ion, cyanide, carbonate, and some carbamates have been detected by this means (Figures 3.37 and 3.38). Bipolar pulse conductivity detectors were reported[21] and have been incorporated in the presently available suppressor-type instruments. Ultraviolet detection has been successful with a variety of ions[22,23] and electrochemical detectors are used to detect a number of partially ionized species.[24] This detector is installed between the separator and suppressor column and can be used with normal conductivity detection with only a minor eluent change (Figures 3.39 and 3.40).

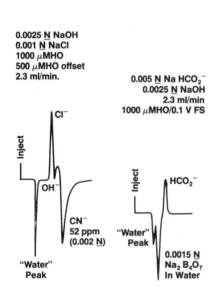

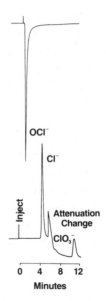

Figure 3.37 and 3.38. Resistivity detection of CN^- and $B_4O_7^{2-}$. Reprinted with permission of *Ann Arbor Science*.[19]

Figure 3.39. Fresh commercial bleach. Copyright by Dionex Corp. and reprinted by permission of copyright owner.

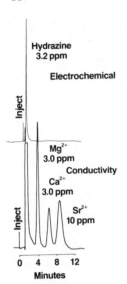

Figure 3.40. Determination of hydrazine. Copyright by Dionex Corp. and reprinted by permission of copyright owner.

3.11 CHROMATOGRAM INTERPRETATION AND QUANTITATION*

Most ion chromatograms are quite simple to interpret and translate into quantitative values for the separated ions. Typically, there are seven or fewer peaks in an ion chromatogram, due in large part to inherent detector limitations and in part to the characteristics of the column packings. Although ions with pK_a or pK_b greater than seven may be eluted under IC conditions, they cannot be detected by conductivity detection. Unless they are at high concentrations the poor ionization of these species results in weak signals. Another factor is the specificity of the resin for some ions. Not all ions can be eluted or separated by one set of operating conditions. Potential peak-producing species are often eliminated by the suppressor columns; for example, cationic species do not create peaks because these ions are removed by the suppressor column in anion analyses. A third factor in the small number of peaks in an ion chromatogram is that the conductivity response in nonionic species is not noticeably large. This means that alcohols, ketones, and hydrocarbons are not detected by IC.

* Much of the information in this section was adapted from E. L. Johnson and R. Stevenson, *Basic Liquid Chromatography*, Varian Associates, Palo Alto, California, 1978, pp. 223–243; see also L. R. Snyder and J. J. Kirkland, *Introduction to Modern Liquid Chromatography*, 2nd ed., Wiley, New York, 1979, pp. 547–574.

In any ion chromatogram a minimum of two types of measurement must be made to translate the peaks into useable data. Each peak's retention time and peak height or area parameters must be determined.

Retention times are used to identify separated ions. The elapsed time from the point at which the eluent is first diverted through the sample loop ("inject") or from a void volume peak to the maximum point of any later eluted peak is the retention time. The retention time of a peak is *not* an absolute identification of an ion. Rather it should be taken as an indication that a certain ion *may* be present. Often more than one species will elute in any given time window. These nearly coeluting peaks may or may not be resolved, depending on the resolution capability of the column set. Knowledge of the source of the sample will help to identify species that may be represented in an ion chromatogram.

After the initial measurements have been made on a chromatogram the retention times are used to identify the eluted ions. In some cases there are peaks whose retention times do not correspond to known ions. Several methods are available to identify unknown peaks. Knowledge of the ion content of the sample is always helpful. The ion chromatography retention tables in the Appendix can also tentatively identify the ions that may be present in a IC peak. The most frequently used method of peak identification compares the chromatograms for unknown samples with artificially prepared standards. Care should be taken to obtain the highest purity chemicals available for standard preparation. Even reagent grade or ultrapure chemicals contain impurities that may be detected by IC and may be mistaken for the ions of interest. In one popular method of confirming a peak's identity the peak is trapped after detection and classical ion tests are run on the trapped material. These classical tests often involve wet chemical or spectroscopic methods. The suppressor column effluent must be considered when a confirming test is run. A fourth way of identifying a peak is to do a coupled chromatographic analysis of the sample. The various eluted ions do not usually exhibit the same retention behavior in the ICE system as they do in the anion IC system (Figure 3.41). Finally, other detectors or flow-injection modifications of the system may provide additional confirmation of the presence of a particular species or identify an unknown.

Three other methods have been used to identify the eluted species: (1) chemical alteration of the sample to remove interferences (i.e. precipitation of bromide ion with Ag_2O or an Ag^+ form column to identify NO_3^- or SO_3^{2-} peaks specifically), (2) spiking the sample with the ion of

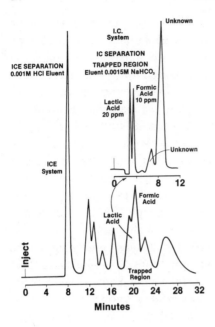

Figure 3.41.   ICE elution followed by weak-eluent IC elution. Copyright by Dionex Corp. and reprinted by permission of copyright owner.

interest from a pure standard, and (3) use of different IC eluents to obtain different characteristic elution profiles. Each of these methods requires extra handling of the sample, which may also introduce errors or additional ions that had not been present. Recovery for the ions of interest must be carefully studied for auxiliary chemical treatments.

The second measurement is a determination of peak height or area to obtain a numerical ion concentration. The vertical distance from the peak maximum to the baseline is the peak height. This distance must be determined by the analyst or data system. To measure peak heights properly chromatogram baselines must be accurately determined. In the most commonly used method in IC a line is drawn to connect a stable region before the peak or peaks of interest with a stable region after these peaks elute. If the peak elutes or another tailing peak (e.g. SO_4^{2-} in high Cl^- concentrations), a curve or preferably a line is drawn to establish the baseline (Figure 3.42). The best method of baseline determination is usually left to the analyst, who should proceed empirically.

Peak-area measurement usually combines peak-height and peak-width measurements. In one popular method the product of peak height (H) and the full width (W) at half-maximum (or FWHM) are used to calculate area (Figure 3.43). A second method (Figure 3.44) is to triangulate the peak by drawing tangential lines for the sides of the peak. The point of inter-

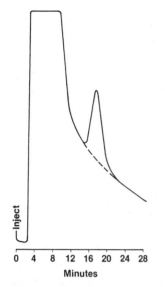

0 4 8 12 16 20 24 28

Minutes

Figure 3.42. Tailing peak baseline determination.

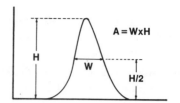

Figure 3.43. Determination of area (FWHM).

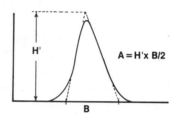

Figure 3.44. Determination of area (tangent and baseline). Copyright by Dionex Corp. and reprinted by permission of copyright owner.

section of the two tangent lines is used as peak height (H'). The baseline width B is measured between the tangent lines. The product of the peak height times half the baseline width is the calculated area.

Another method of peak-area or peak-height measurement uses an electronic data system. Some of these systems have built-in recorders and normally report retention time, peak height, and/or area and can usually calculate quantitative values automatically. Several integrator parameters must be optimized by the analyst before reliable data can be obtained. This is usually done empirically by referring to the integrator manufacturer's instructions (unfortunately too often these are obtuse) and then experimenting with several standards that closely resemble the samples.

Once the peak identity is established quantitative values for the eluted ion must be calculated. A number of different methods have been used in IC to determine numerical values for eluted species:

1. Calibration plot for each eluted ion.
2. External standard from one synthetic standard.
3. Internal standard.
4. Method of additions.

Calibration plots are used when wide ranges of ions are expected in the samples. A high concentration standard (usually 1000 μg/mL of the ion of interest) is prepared from *dried* reagent-grade chemicals. This 1000-ppm standard is usually stable for several months. The lower concentration standards should be prepared weekly or more often. Appropriate blanks for the eluent or H_2O should be run each time a new standard is prepared.

A calibration plot is established by running a series of appropriate standards on the ion chromatograph. The standards are measured for peak height/area and retention time. The peak response (height or area times the scale setting) is plotted against concentration. It is important to note that the individual ion response is different for each ionic species. Thus each ion will require its own calibration curve. Figure 3.45 is a typical calibration plot. As in most curves, this plot is reasonably linear over a fairly wide range. Experience has shown that once an IC calibration curve is determined the day-to-day working curve has the same slope as the calibration plot but the intercept of the curve will change slightly with laboratory temperature. Thus once a multipoint calibration plot is generated it normally needs only to be verified by a several-point working curve for daily instrument use. Of course, this depends on the condition of the instrument and column set. The ends of the calibration plot are sometimes nonlinear. Interaction between the suppressor column effluent and the eluted ion at low concentrations may be the cause of nonlinearity at a low concentration range. Also, the calibration plot does not typically pass through zero. At the other end of the calibration curve the nonlinearity results from the high ion concentration in the cell. The dilute solution approximation that linearly equates conductance to ion concentration no longer holds true when high concentrations are passed through the cell. Thus ion concentrations in excess of a few hundred ppm begin to show nonlinearity. Although this region of the calibration plot is slightly nonlinear, it may still be used for reasonable estimation of ion concentrations.

Another type of external standard calibration is done with a prepared single synthetic standard that closely approximates the sample. Knowledge of the sample composition is critical because the standard is prepared by estimation of the ion concentrations in the unknown. The peak heights for the standard are compared by ratio to the sample peak heights to obtain final quantitative values for the ions of interest. Figure 3.46 is a typical chromatogram of this type of quantitative ion determination.

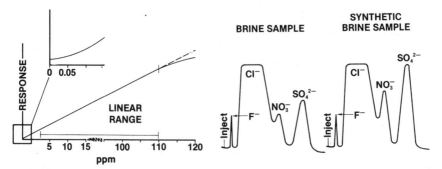

Figure 3.45. Typical IC calibration curve.   Figure 3.46. Use of synthetic standard.

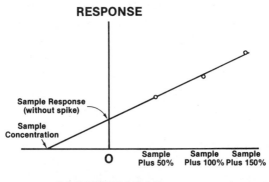

Figure 3.47. Method of standard additions. Copyright by Dionex Corp. and reprinted by permission of copyright owner.

Two other methods are used occasionally in IC analyses. The first is standard addition. This quantitation method requires several injections of sample and is needed only when the sample is quite complex. Initially, the sample is run with a very small volume spike of deionized water. Three more samples are spiked with the same volume of varying concentrations of the standard. The spikes range from 50% more than the estimated ion value to 150% more than the value. These three sample/standards are run on the IC and then plotted (Figure 3.47). The line is extrapolated to give the ion concentration in the original sample. Another method used occasionally in IC is internal standardization which also requires changing the ionic composition of the sample. An ion not present in the sample is chosen for peak-height comparisons. The peak heights or areas of the ions of interest are compared by ratio to the internal standard ion for an artificial standard. The sample is then spiked with the internal standard and the relative response factor is used to calculate the concentration of the ion of interest.

Once the data have been acquired and reduced it must be placed in a meaningful context. Several terms are used to describe data and to put it into perspective. The accuracy of a method is the difference between a measured value and the true value for a sample constituent. This should not be confused with data comparisons between two instrumental methods such as IC and wet chemistry. True values for nonsynthetic or prepared samples are difficult to obtain. Usually the only way to obtain a true value is in a long-term study of a sample by many diverse analytical methods. Because all instrumental methods contain errors, analysis of a sample by only one method will give only an approximation of the true value for species concentration. Accuracy can be estimated in recovery studies done with spiked nonsynthetic samples. Table 3.2 lists data for the recovery of a variety of ions run in water samples.[25] Inevitably, analytical methods are compared under the guise of "accuracy." Instrument manufacturers and their customers alike indulge in this insanity. Ultimately all that can be said of data obtained by IC is that round-robin and individual studies have shown that IC produces results comparable to values obtained by a variety of other instrumental methods.

Precision and repeatability, often used interchangeably, need more careful definition because they do not have the same meaning. Precision is the expression of the differences of measured values for a given set of data. It is described as range, standard deviation [often as percent relative standard deviation (% RSD)], or average deviation. Repeatability is the precision of a given operator working with a given instrument and with a given analytical method. Reproducibility is the precision of different operators working with the same analytical method but often on different instruments. Precision of the IC is generally better than 3% and reproducibility is better than 5%. IC will give results of ion analyses that compare quite favorably with other methods in terms of "accuracy," intermethod comparisons, precision, and reproducibility.

3.12 CHROMATOGRAM ANOMALIES

Although most ion chromatograms are quite simple, in some instances artifacts complicate interpretation of the data. A variety of methods exists

Table 3.2   Recovery Data On Precipitation Samples Results in Milligrams per Liter

Sample No.	Constituent	Present	Added	Found	Percent Recovery
1	F	0.038	0.04	0.08	102
	Cl	0.294	0.3	0.61	103
	NO_3-N	0.56	0.6	1.14	98
	SO_4	10.54	10.0	19.7	96
2	F	0.052	0.05	0.105	103
	Cl	0.385	0.4	0.77	98
	NO_3-N	0.618	0.6	1.21	99
	SO_4	6.74	7.0	14.2	103
3	F	0.043	0.04	0.08	96
	Cl	0.304	0.3	0.62	103
	NO_3-N	0.475	0.5	0.95	97
	SO_4	5.70	6.0	11.2	96
4	F	0.063	0.06	0.125	102
	Cl	0.945	1.0	1.99	102
	PO_4-P	0.057	0.06	0.12	102
	NO_3-N	1.39	1.5	2.95	102
	SO_4	7.60	7.0	15.1	103
5	F	0.030	0.027	0.057	100
	Cl	0.95	1.0	2.0	102
	NO_3-N	0.428	0.4	0.81	98
	SO_4	3.52	3.73	7.2	99
6	F	0.057	0.053	0.11	100
	Cl	0.55	0.53	1.04	96
	NO_3-N	1.76	1.75	3.45	98
	SO_4	9.12	9.33	18.0	98
7	F	0.029	0.04	0.07	101
	Cl	0.51	0.5	0.98	97
	NO_3-N	1.31	1.5	2.81	100
	SO_4	7.98	7.0	15.5	103
8	F	0.024	0.025	0.05	102
	Cl	0.314	0.3	0.60	98
	NO_3-N	0.88	1.0	1.8	96
	SO_4	7.32	7.0	14.4	100
9	F	0.074	0.07	0.143	99
	Cl	0.332	0.5	0.82	98
	NO_3-N	0.76	0.75	1.48	98
	SO_4	8.3	7.0	15.8	103
10	F	0.057	0.06	0.115	98
	Cl	0.446	0.5	0.95	100
	NO_3-N	0.774	0.75	1.48	97
	SO_4	11.3	7.0	18.8	103

to minimize artifacts, but these measures must often be used *before* the sample is injected. Knowledge of the sources of the artifacts is useful in anticipating trouble.

One common problem in anion chromatograms is the "water dip." There are actually two negative dips in the baseline, both early in the chromatogram. The first small sharp dip occurs just before the F^- peak. This is the void volume indication of the column set. It is most often seen when operating at low conductivity meter readings or when clean samples are being analyzed. A second depression in the chromatogram baseline closely follows the void volume indication. This second negative response is deeper and much broader and has been termed the "water or carbonate dip." A variety of sources for the dip has been postulated, among which are eluent dilution and the elution of pure water or low concentrations of OH^- ion. Whatever the cause, water-based samples have the largest dips. The shape of the dip depends on the volume of sample injected, on the volume of resin in the suppressor column, and on the amount of regenerated suppressor resin remaining. The last is important because the dip shape changes as the suppressor resin is expended. This changing shape alters the peak height for rapidly eluting species. The amount of negative depression in the chromatogram is also a direct function of offset required for the eluent. Lower strength eluents or those that are converted to H_2O in the suppressor show small dips, whereas high-strength eluents or those that emerge from the suppressor with moderate (10 to 30 μMHO/cm) conductivity have large dips. Thus $NaOH/Na_2B_4O_7$ anion eluents and HCl or HNO_3 cation eluents have low dips, whereas $NaHCO_3/Na_2CO_3$ anion eluents have more pronounced depressions in the baseline.

One way to minimize the dip is to match the sample to the eluent. For standard anion analyses this is most often done by adding sufficient HCO_3^-/CO_3^{2-} ions to the sample to make the concentration of these ions almost equal between sample and eluent. The fluoride and chloride calibration curves will have higher intercepts because two positive peaks coincide with these peaks.

A second way to minimize the effect of the dip is to use a weaker eluent or eluents that form water in the suppressor. The depth of the dip is less because the conductivity offset is lower than with standard eluent. A $0.0015M$ $NaHCO_3$ eluent has about one-third the dip that the $0.003M$ $NaHCO_3/0.0024M$ Na_2CO_3 eluent shows. This weaker eluent is effective in separating Cl^- and NO_2^- from the water dip but not in eluting NO_3^-, SO_4^{2-}, or other moderately retained ions. NaOH-based eluents will show small dips, but again these eluents are not effective in eluting SO_4^{2-}. The high concentrations of NaOH required to elute SO_4^{2-} expend the suppressor column rapidly.

A third alternative is a step gradient anion elution. The gradient sequence starts with a weak CO_3^{2-} eluent to elute the weakly held ions and then progresses to successively more concentrated eluents to elute moderately and strongly held ions. The use of a single ionic eluting species at different concentrations is necessary to minimize equilibration time in returning from the strongest to the weakest eluent.

Another difficulty observed in some ion chromatograms occurs in shifts in retention time. These peak shifts are observed with concentration increases, interaction between the sample and the eluent, and temperature. The potential problem of peak shift can be observed as far back as the earliest publication on IC. Small et al. characterized the retention behavior for numerous aliphatic amines.[26] The pattern of shorter elution time at higher ion concentrations can be observed in this work. For anions it appears that some species (NO_2^-, NO_3^-, I^-, and SCN^- as examples) are more prone to peak shift with concentration than other eluted ions. One example of peak shift caused by sample–eluent interaction is in the elution of SO_4^{2-} ion in NaOH solutions. When standard $0.003M$ $NaHCO_3/0.0024M$ Na_2CO_3 eluent is used the OH^- in the sample alters the eluent by converting some of the HCO_3^- to CO_3^{2-}, a much stronger eluting ion. This causes the SO_4^{2-} to elute faster.[27] One ion exhibits odd behavior with temperature. Phosphate ion elutes later in the chromatogram as temperature rises, whereas most other peaks elute earlier.

Several solutions are available for peak-shift problems. The first is dilution of the sample in eluent, which is preferred if an interaction between sample ions and eluent is suspected. At the lower concentrations there is less likelihood of overloading the separator column. The second method spikes the sample by adding a small volume of a standard solution to the sample and then rerunning the chromatogram. This helps to verify the identity of a shifted peak and can be used to determine the concentration of the ion of interest if the amount of the spike is known.

Apparently peak shifts are sometimes caused by high concentration constituents in the sample which may tie up so much of the limited separator column capacity that other ions cannot compete effectively for the exchange capacity. Also, the high concentration ion may act as an eluent for the other ions. This problem occurs most often when the major constituent has a high affinity for the resin (i.e., PO_4^{3-} and SO_4^{2-}) rather than a low affinity (e.g. Cl^-). The best approach to a minimization of this problem is to dilute or pretreat the sample to keep the total quantity of ions injected onto the separator column reasonably low.

Another artifact seen in ion chromatograms is a drifting or shifting baseline. The slow elution of strongly retained ions and room temperature changes are the main causes of baseline drift. Ions that have high affinities for the separator resin elute slowly. This can produce a domelike baseline disturbance. Excessive temperature compensation on the conductivity board will also cause drift which generally appears as a long-term sine wave. Careful adjustment of the amount of temperature compensation will produce flat baselines if room temperature fluctuations are not too severe. Thermostating column sets and detector will also minimize temperature effects.

Shifts in baseline are caused by sample loading problems or eluent changes. If the separator/suppressor column is overloaded by a high ion concentration, a slow return to the original baseline will result. Figure 3.48 is an example. The excessive Cl^- ion concentration (about 60,000 ppm) causes a 2-h tailing peak before the original baseline is restored.

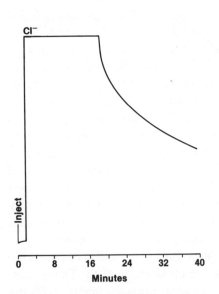

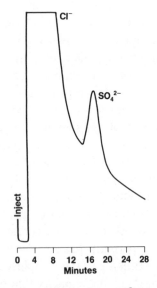

Figure 3.48. Direct injection of 10 ppm SO_4^{2-} in 10% NaCl brine. Copyright by Dionex Corp. and reprinted by permission of copyright owner.

Figure 3.49. Analysis of 1% SO_4^{2-} in 1% NaCl. Copyright by Dionex Corp. and reprinted by permission of copyright owner.

The sample should be diluted before reinjection; Figure 3.49 shows a more rapid return to baseline when the Cl^- concentration is reduced to about 6000 ppm.

Another type of shift is due to the use of eluent step gradient elution. As the concentration of CO_3^{2-} is increased in anion step gradient elution the H_2CO_3 background also increases. The amount of offset to get an electrically zero baseline must be reset for each eluent change. The levels of offset are most often established as the step gradient separation is optimized.

The suppressor resin interacts with some ionic species while performing eluent suppression and conversion of the sample ions to a common form; for example, in the anion system the suppressor resin interacts with weak acid species. Similarly, the cation suppressor resin interacts with ammonium ion and some quaternary ammonium ions. The interaction mechanisms are the same as those in IE and ICE separations. Species move through the suppressor resin at different velocities when they are dissociated (ionic) than when they are nondissociated (molecular). When the species are not dissociated they can penetrate the resin bead and be retained. The strong acid species (e.g. F^-, Cl^-, and SO_4^{2-}) are separated by Donnan exclusion and pass right through the column. These ions show up as the void volume peak. Elution of the organic acids is altered by the partition of these species in the resin. This suppressor interaction of weak acids can be minimized by using small suppressor resin beds or sometimes by adding alcohol or other organic solvents to the eluent. Figure 3.50

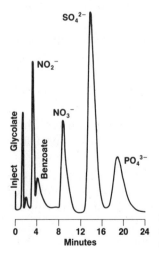

Figure 3.50. Anions in ethylene glycol. Copyright by Dionex Corp. and reprinted by permission of copyright owner.

shows a separation of benzoate ion in ethylene glycol. With a large 6 × 250 mm glass anion suppressor the benzoate peak is broad and not easily quantitated. When a 3 × 250 mm column is used the benzoate ion forms a sharp peak that is well separated and easily quantitated along with the other ions present. Use of the hollow-fiber membrane suppressor should also minimize this type of interaction problem.

It can truthfully be said that the number of ions separable by IC has grown significantly in the last five years. Along with this development the IC method has rapidly increased its capabilities and sophistication. As more rapid specific analyses are required, instrumental methods will evolve to meet these needs. The practice of IC is expected to change considerably in the next several years.

NOTES

1. W. Rich, F. Smith Jr., L. McNeil, and T. Sidebottom, "Ion Exclusion Coupled to Ion Chromatography: Instrumentation and Application," in J. D. Mulik and E. Sawicki, Eds., *Ion Chromatographic Analysis of Environmental Pollutants,* Vol. 2, Ann Arbor Science, Ann Arbor, Michigan, 1979, pp. 17–29.

2. *Ion Chromatography Training Course,* Dionex Corporation, 1978.

3. W. F. Koch, "Complications in the Determination of Nitrite by Ion Chromatography," *Anal. Chem.,* 51(9), 1571–1573 (1979).

4. Anonymous, "Use of Small Suppressors," *IC Exchange,* Dionex Corporation, 1980.

5. T. S. Stevens, V. T. Turkelson, W. R. Albe, "Determination of Anions in Boiler Blow-down Water with Ion Chromatography," *Anal. Chem.,* 49(8), 1176–1178 (1977).

6. E. L. Johnson, Dionex Corporation, Sunnyvale, California, private communication.

7. F. C. Smith, Jr., "Step Gradient Elution of Anions with Ion Chromatography," presented at the Dionex Symposium on Ion Chromatography, Sunnyvale, California, June 1978.

8. F. C. Smith, Jr., V. T. Smith, and J. K. Ung, "Optimized Analysis of Brine and Caustic Samples by Ion Chromatography," presented at the Pittsburgh Conference on Analytical Chemistry and Applied Spectroscopy, Atlantic City, New Jersey, March 1980.

9. W. Dahl and P. McCullough, Monsanto Corporation, St. Louis, Missouri, private communication.

10. Dionex Corporation, Sunnyvale, California, private communication.

11. R. A. Wetzel, C. L. Anderson, H. Schleicher, and G. D. Crook, "Determination of Trace Level Ions by Ion Chromatography with Concentrator Columns," *Anal. Chem.,* **51**(9), 1532–1535 (1979).

12. R. K. Pinschmidt and T. P. Katrinak, "Analysis of Chloride and Other Ions in Acetic Acid," in J. D. Mulik and E. Sawicki, Eds., *Ion Chromatographic Analysis of Environmental Pollutants,* Vol. 2, Ann Arbor Science, Ann Arbor, Michigan, 1979, pp. 31–40.

13. B. Hickam, J. Bellows, D. Pensenstadtler, and S. Peterson, "Determination of Steam Purity in Turbines and Power Plants," Westinghouse Corporation, 1979.

14. R. K. Pinschmidt and T. P. Katrinak, "Analysis of Chloride and Other Ions in Acetic Acid," in J. D. Mulik and E. Sawicki, Eds., *Ion Chromatographic Analysis of Environmental Pollutants,* Vol. 2, Ann Arbor Science, Michigan, 1979. pp. 31–40.

15. F. C. Smith, Jr., Microsensor Technology Inc., Fremont, California, unpublished work.

16. S. A. Bouyoucos, "Determination of Ammonia and Methylamines in Aqueous Solutions by Ion Chromatography," *Anal. Chem.,* **49**(3), 401–403 (1977).

17. F. C. Smith, Jr., Microsensor Technology, Inc. Fremont, California, unpublished work.

18. R. C. Chang, General Electric, Mt. Vernon, Indiana, unpublished work.

19. R. K. Pinschmidt, Jr., "Ion Chromatographic Analysis of Weak Acids Using Resistivity Detection," in J. D. Mulik and E. Sawicki, Eds., *Ion Chromatographic Analysis of Environmental Pollutants,* Vol. 2, Ann Arbor Science, Ann Arbor, Michigan 1979. pp. 41–50.

20. R. K. Pinschmidt, Jr., and M. J. Fasolka, "Resistivity Detection of Weak Acids: Carbamates and Carbonates." presented at the Rocky Mountain Conference, Ion Chromatography Symposium, Denver, Colorado, August 1981.

21. J. M. Keller, "Bipolar-pulse Conductivity Detector for Ion Chromatography," *Anal. Chem.,* **53**(2), 344–345 (1981).

22. R. J. Williams, "Application of Ion Chromatography to the Analysis of Organic Sulfur Compounds," presented at the Rocky Mountain Conference on Analytical Chemistry, Denver, Colorado, August 1980.

23. E. Cathers, and E. L. Johnson, "Applications of Simultaneous Detection in IC," presented at the Rocky Mountain Conference, Symposium on Ion Chromatography, Denver, Colorado, August, 1981.

24. Dionex Application Note No. 29.

25. M. J. Fishman and G. Pyen, *Determination of Selected Anions in Water by Ion Chromatography,* United States Geological Survey, Water Resources Investigations, Publication USGS/WRI—79–101 (1979).

26. H. Small, T. S. Stevens, and W. C. Bauman, "Novel Ion Exchange Chromatographic Method Using Conductimetric Detection," *Anal. Chem.,* **47**(11), 1801–1809 (1975).

27. F. C. Smith, Jr., Microsensor Technology Inc., Fremont, California, unpublished work.

4

Nonsuppressed
Ion Chromatography

The term "ion chromatography" was once applied only to ion-exchange separations of simple ions in which a suppressor column/conductivity cell was used as detector. Although most IC analyses are still done with this detection system, one of the system's important limitations leaves many common ions (CN^-, S^{2-}, etc.) unanalyzable by IC. The scope of ion chromatography has recently been broadened by the introduction of complimentary methods of analysis. For simple ions with pK less than 5 nonsuppressed IC methods have been developed to compete directly with suppressor-type IC. For ions with pK greater than 5 the application of new detectors has permitted the use of IC in the determination of weaker acids and bases (which are often carbon-based).

This chapter covers the resins and methods of competing single-column IC separation systems for strongly acidic or basic ions and explores some of the applications of single-column methods that involve new detectors. Some novel "hyphenated" separation–detection methods which extend the range of ion chromatography substantially are also covered. Because these methods are still quite new as this text is prepared, the body of information from which to pull examples is naturally not so great as it is for suppressor-type IC.

4.1 NONSUPPRESSED ION CHROMATOGRAPHIC METHODS FOR IONS WITH pK < 5

The success of suppressor-type IC has led to the introduction of nonsuppressed IC instrumentation by several instrument manufacturers, with a concurrent increase in competition.[1] Single-column methods have been developed to complement suppressor type IC and to circumvent the Dow patents. A few papers on nonsuppressed IC have been published.[2-7] Sections 4.2 and 4.3 detal current practices in nonsuppressed IC detection of strongly ionized species.

4.2 RESINS USED IN NONSUPPRESSED IC FOR STRONG IONS

The most popular nonsuppressed IC methods have two main variations, both based only on the use of a separating column and conductivity detector. The first method utilizes a silica-based pellicular ion exchanger for anion separations. The second method uses an aminated macroporous styrene/divinylbenzene (S/DVB) resin for the separation of anions. For cations several column packings have been tried, but a low-capacity sulfonated gel-type S/DVB resin seems to work best. Details of the styrene/ divinylbenzene resins are given in Chapter 5.

Silica-based anion separation methods are used in ion chromatographs supplied by Wescan, Hewlett-Packard and in the HPLC add-on components supplied by The Separations Group. The pellicular anion exchange resin has aminated active ion exchange capacity attached to a silica substrate by Si—C bonding. The resulting resin has a low 0.1 meq/g capacity and is sold under the Vydac trade name. Like other silica-based packings, some factors should be considered when these resins are used for ion analyses. The column packings should not be exposed to samples or eluents whose pH is less than 2 or greater than 7. Severe degradation of the packing may occur if these limits are exceeded. Also, high concentrations of fluoride and hydroxide ions should be avoided. Samples whose composition would normally cause problems in the separator packing can often be run if they are diluted with eluent. This balances the pH of the sample with the eluent pH. Some loss of capacity in the silica packing is encountered even when these precautions are routinely observed.[8]

The silica anion exchanger has substantially different selectivity for anions than the S/DVB anion separator resins used in suppressor-type IC (Chapter 5). Figure 4.1 shows that the separation between bromide and nitrate is quite large in silica columns. These two anions nearly coelute in suppressor-type anion IC. Note also that nitrite and bromide now nearly coelute with the silica packing. One resolution problem is traded for another, as is usually the case. Another difference is in the elution order of sulfur oxides. Figure 4.2 shows the elution of sulfate, thiosulfate, and sulfite (in that order). The elution order for these three species in suppressor-type IC is sulfite, sulfate, and then thiosulfate. These differences in selectivity can be used advantageously when a relatively large amount of one ion (the last eluted) is present and the other two are present in a lower concentration.

The second variation of nonsuppressed anion IC utilizes a macroporous styrene/divinylbenzene resin for ion separations. This resin is not yet commercially available, although it or one of its near relatives should be marketed in the near future. It has been prepared from ground and sieved macroporous XAD-1 resin (Rohm and Haas, Philadelphia). Quaternary

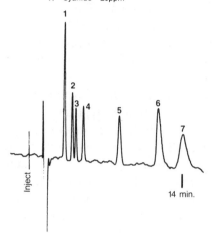

Column:	VYDAC 302IC4.6
Detector:	VYDAC 6000 CD—Conductivity Detector
Solvent:	4mM Phthalic acid, pH 4.5 c̄ NaBorate
Flow Rate:	2ml/min
Injection:	200 microliters of:
1.	Chloride—10ppm
2.	Nitrite—10ppm
3.	Bromide—10ppm
4.	Nitrate—10ppm
5.	Sulfate—10ppm
6.	Thiosulfate—20ppm
7.	Cyanide—20ppm

Figure 4.1. Separation of common anions by nonsuppressed IC. Courtesy of Wescan Corp.

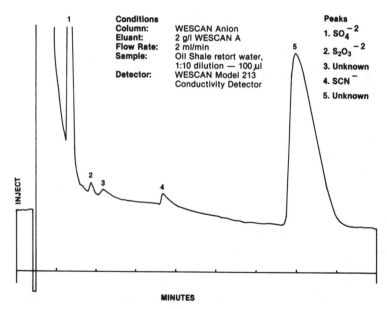

Conditions

Column:	WESCAN Anion
Eluant:	2 g/l WESCAN A
Flow Rate:	2 ml/min
Sample:	Oil Shale retort water, 1:10 dilution — 100 μl
Detector:	WESCAN Model 213 Conductivity Detector

Peaks

1. SO_4^{-2}
2. $S_2O_3^{-2}$
3. Unknown
4. SCN^-
5. Unknown

Figure 4.2. Elution of sulfur oxides. Courtesy of Wescan Corp.

ammonium active capacity is introduced by reacting the polymer with chloromethyl methyl ether and then aminating it with trimethylamine. Great care must be taken in the chloromethylation step because chloromethyl methyl ether is a known human carcinogen. Details of preparation are available.[9,10] The final product has an ion exchange capacity between 0.007 and 0.07 meq/g. The resin is then packed into a column (3 × 500 mm). Figure 4.3 shows separations of common anions done with this resin. Once again differences in selectivity from gel-type S/DVB resins can be observed. The elution order of I^-, SCN^- and SO_4^{2-} is quite different from the suppressor-type IC elution of these ions.

Nonsuppressed cation separation IC methods have been reported,[11] but this method has not been adopted as a common IC application. Separations are done on resins that are similar to those used in suppressor-type IC. A low cross-linked gel-type S/DVB resin is sulfonated with hot H_2SO_4 or with chlorosulfonic acid. The reaction is quenched with water and the resulting resin is packed into a column. Mineral acids and amine eluents are used to elute monovalent and divalent cations (Figures 4.4 and 4.5). One unusual feature of the cation IC system is the detection of the eluted species as negative peaks, which result from the separated ions having lower conductance than the high conductivity background of the eluents.

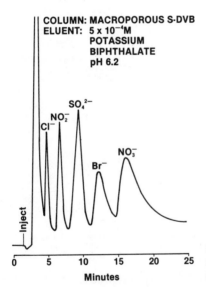

COLUMN: MACROPOROUS S-DVB
ELUENT: 5×10^{-4}M
POTASSIUM
BIPHTHALATE
pH 6.2

SO_4^{2-}
NO_2^-
Cl^-
Br^-
NO_3^-
Inject

0   5   10   15   20   25
Minutes

Figure 4.3. Macroporous S-DVB separation of anions. Reprinted from the *Journal of Chromatography* with permission.[10]

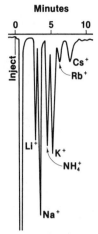

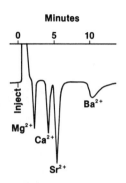

Figure 4.4. Monovalent cations by single-column IC. Reprinted from *Analytical Chemistry* with permission.[11] Copyright 1980 American Chemical Society.

Figure 4.5. Divalent cations by nonsuppressed IC. Reprinted from *Analytical Chemistry* with permission.[11] Copyright 1980 American Chemical Society.

Non-ion exchange anion separations can be done on a single-column ion chromatographs. Anions with pK_a between 2 and 5 have been separated by a cation exchange resin in a separation exactly like the anion ion chromatography exclusion (ICE). This separation process has been termed ion moderated partition by Bio-Rad of Richmond, California. No suppressor column is used for these separations. A mineral acid eluent (sometimes combined with an organic solvent such as acetonitrile or methanol) maximizes the separation of a variety of species. Figure 4.6 shows the separation of some organic acids and sugars with a low-wavelength UV detector. Conductance detection with these separation methods has not been widespread.

4.3 ELUENTS COMMONLY USED IN NONSUPPRESSED IC FOR STRONG IONS

One major consideration in nonsuppressed IC is the choice of eluting ions. Because a low conductivity background is desirable for the detection of trace ion concentrations, a high-affinity eluting ion at low concentration is the usual choice. Both the pH and the ionic strength of the eluent must be closely controlled when the eluent is prepared.

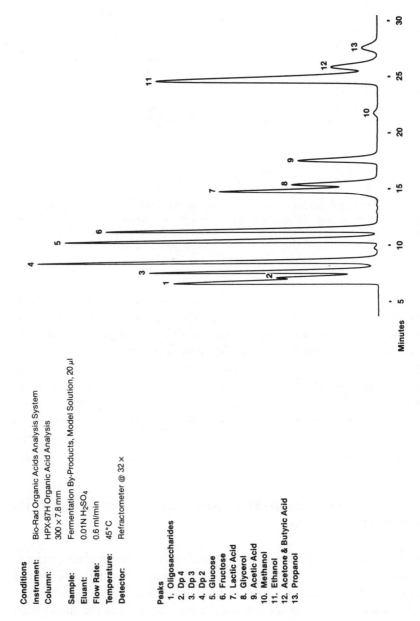

82

Conditions

Instrument:	Bio-Rad Organic Acids Analysis System
Column:	HPX-87H Organic Acid Analysis 300 × 7.8 mm
Sample:	Fermentation By-Products, Model Solution, 20 μl
Eluant:	0.01N H_2SO_4
Flow Rate:	0.6 ml/min
Temperature:	45°C
Detector:	Refractometer @ 32×

Peaks

1. Oligosaccharides
2. Dp 4
3. Dp 3
4. Dp 2
5. Glucose
6. Fructose
7. Lactic Acid
8. Glycerol
9. Acetic Acid
10. Methanol
11. Ethanol
12. Acetone & Butyric Acid
13. Propanol

Minutes

Figure 4.6.   IMP separation of organic acids and sugars. Courtesy of Bio-Rad Corp.

Aromatic acid salts, which are one class of compound that exhibits desirable characteristics as eluents, are widely used in nonsuppressed IC. Benzoate and phthalate ions have been applied as eluting ions in the separation of a variety of anions and are used in the 10^{-3} to $10^{-5}M$ concentration range to elute a number of ions. The eluents are pH-adjusted with sodium borate or pyridine. Conductivity backgrounds range from 20 to 250 μMHO/cm. Weakly retained ions are eluted by low concentrations of the high-affinity eluents or by eluents prepared from aliphatic organic acids such as acetic, formic, and glutamic. Acids with higher affinity (e.g., citrate) elute the more strongly held species.

Anion eluents are adjusted to specified pH when they are prepared. The most widely used pH range is normally 5 to 7; pH between 6 and 6.5 are the most frequently reported. Distilled or deionized water and high purity chemicals should be used in eluent preparation. Guidelines for eluent selection and preparation have been developed and are available from instrument manufacturers.

Cation separations by nonsuppressed IC are done with nitric acid for monovalent cations and ethylenediammonium dinitrate for divalent cations. These eluents are also used in the 10^{-3} to $10^{-5}M$ range. Their conductivity backgrounds are 1000 μMHO/cm or more. Separated ions are detected as negative peaks.

Because single-column nonsuppressed IC methods for common anions are recent developments, relatively few unique applications have been done. Most applications tend to parallel suppressor-type IC. Existing applications sheets available for nonsuppressed IC include (among others) drinking water, industrial formulations, Bayer and Kraft black liquors, and ultratrace analysis.[12] In general, nonsuppressed IC has been used for samples in which the matrices are not especially complex (drinking waters, ultrapure waters, etc.) or in which sample constituents are high enough in concentration to be easily detected by conductance measurements after the samples are substantially diluted. The limitations from sample and eluent pH and sample composition are problems that are not easily solved for the silica resin. These problems do not affect macroporous S/DVB resin. Development of new resins and eluents should help to eliminate these difficulties. The range of applications of nonsuppressed IC will probably be as broad as that of suppressor-type ion chromatography.

4.4 IC METHODS FOR THE DETERMINATION OF WEAKLY IONIZED SPECIES (pK RANGE 5 TO 13)

The use of conductivity detection in ion chromatography imposes some severe limitations on the types of ion that can be detected. This detection system will, in general, detect only those ions whose pK is less than 5. For many ionic species with pK greater than 5 there is insufficient conductivity from these ions as they pass through the cell to give adequate signals for trace detection. This appears to be especially true for suppressor-type IC, where the ions are often compounded by conversion to their H^+ analogs for anions or their OH^- form for cations. The former makes weak acid of eluted ions such as CN^- and S^{2-}; the latter appears to precipitate transition metals in the suppressor column. However, cyanide ion has been reported to be eluted and detected by nonsuppressed IC.[13]

It is limiting to apply the term ion chromatography only to conductivity detection. Electrochemical and UV detectors have been used in HPLC for many years. The electrochemical detector system in the Dionex suppressor-type ion chromatograph is really a nonsuppressed IC method because the detector lies between the separator and suppressor columns. Thus the term ion chromatography has already been expanded.

Probably the most important new detector is the electrochemical cell/potentiostat which appears to be capable of directly detecting electroactive ions in the 1 to 10 μg/L range. At least three companies manufacture an electrochemical detector (first marketed by Bioanalytical Systems, Lafayette, Indiana, and also offered by Dionex, Sunnyvale, California, and PAR, Princeton, New Jersey). Table 4.1 is a list of some of the weakly ionized species that can be determined by electrochemical detection after separation on IC columns. Table 4.2 is a list of commonly used elution

Table 4.1. Some Inorganic Ions Determined Electrochemically

CN^-	ClO_3^-
S^{2-}	SCN^-
Br^-	I^-
$N_2H_5^+$	$S_2O_3^{2-}$
ClO^-	

Table 4.2. Elution Conditions Used for Electroactive Ions with Suppressor-Type IC Separating Columns

Ion	Eluent
Standard anion, CN^-	$0.002M$ Na_2CO_3 with 0.75 mL/L ethylenediamine (pH 11.3), 5.0% isopropanol[a]
S^{2-}	$0.0024M$ Na_2CO_3, $0.0030M$ $NaHCO_3$
Br^-, I^-, SCN^-, $S_2O_3^{2-}$	$0.0024M$ Na_2CO_3, $0.0030M$ $NaHCO_3$
I^-, SCN^-, $S_2O_3^{2-}$	$0.005M$ Na_2CO_3 with $0.00075M$ p-cyanophenol/anion precolumn
ClO^-, ClO_3^-	$0.002M$ Na_2CO_3/Flow injection
$N_2H_5^+$	$0.005M$ lysine hydrochloride, $0.003M$ HCl/Standard cation column

[a] The ethylenediamine is strongly retained on the suppressor column. A three-step regeneration is required to restore the suppressor to its original capacity. The column must first be rinsed with $0.1M$ sodium chloride, then with $0.5M$ sulfuric acid, and finally with deionized water.

Table 4.3. Some Organic Anions Determinable by Electrochemical Detection

Hydroquinone	3-Hydroxyphenol
Phenol	4-Hydroxyphenol
2-Chlorophenol	2-Hydroxyphenol
4-Chlorophenol	4-Methoxyphenol
2,6-Dichlorophenol	4-Hydroxybenzaldehyde
2,4-Dichlorophenol	3-Methoxyphenol
2,4,6-Trichlorophenol	2-Methylphenol
2,4,5-Trichlorophenol	4-Methylphenol
2,3,4,6-Tetrachlorophenol	3-Methylphenol
Pentachlorophenol	Bisphenol A
3,4-Dichlorophenol	N-Acetyl-4-aminophenol
2,4-Dichloro-5-methoxyphenol	4-Nitrophenol
2,4-Dibromophenol	2-Hydroxybenzaldehyde
2-Bromophenol	2-Nitrophenol
2,3,5,6-Tetrachlorophenol	4-Nonylphenol
2,3,4,5-Tetrachlorophenol	Ionol
2,4,6-Tribromophenol	2,4-Dinitro-6-S-butylphenol
2,2',6,6'-Tetrabromobisphenol A	2-Phenylphenol
Oxalic acid	4,4'-Dihydroxybiphenyl
Ascorbic acid	4-Phenylphenol
Formic acid	3-Hydroxybenzoic acid
2,4-Dihydroxybenzoic acid	2-Hydroxybenzoic acid
3-Hydroxybenzoic acid	3,5,6-Trichloro-2-pyrimidol

Reference 14.

conditions for the separation of certain simple electroactive anions and cations. Table 4.3 is a list of some of the chlorinated hydrocarbons that can be electrochemically detected.[14] Figures 4.7 and 4.8. are typical chromatograms of electrochemical detection of weak acid ions. Figure 4.9 is an example of the separation and electrochemical-UV detection of chlorinated phenolic species.

Another important detector that is just beginning to be applied in IC is ultraviolet detection. New UV detectors are capable of operating in the 190 to 220 nm range in which some common ions exhibit absorbance.

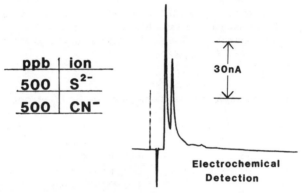

ppb	ion
500	S^{2-}
500	CN^-

Figure 4.7.   Cyanide ion with electrochemical detection. Copyright by Dionex Corp. and reprinted by permission of copyright owner.

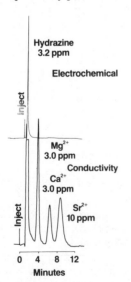

Figure 4.8.   Determination of hydrazine. Copyright by Dionex Corp. and reprinted by permission of copyright owner.

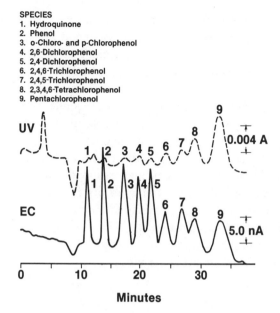

SPECIES
1. Hydroquinone
2. Phenol
3. o-Chloro- and p-Chlorophenol
4. 2,6-Dichlorophenol
5. 2,4-Dichlorophenol
6. 2,4,6-Trichlorophenol
7. 2,4,5-Trichlorophenol
8. 2,3,4,6-Tetrachlorophenol
9. Pentachlorophenol

Figure 4.9.   Separation and detection of chlorinated phenolic compounds. Reprinted from *Analytical Chemistry* with permission.[14] Copyright 1979 American Chemical Society.

Table 4.4 is a list of some of the ions presently determinable by UV after separation by suppressor-type IC methods.[15] Use of the UV detector in-line after the suppressor column makes simultaneous UV/conductivity detection feasible for routine analytical work.

Finally, "hyphenated" analytical methods are now becoming more important in ion chromatography. These hybrid methods are developed

Table 4.4. Anions Determined by Low Wavelength UV Detection

S^{2-}	NO_2^-
SO_3^{2-}	NO_3^-
SCN^-	N_3^-
$S_2O_3^{2-}$	Cl^-
SeO_3^{2-}	Br^-
SeO_4^{2-}	I^-
$SeCN^-$	IO_3^-
AsO_3^{3-}	BrO_3^-
AsO_4^{3-}	ClO_3^-
	ClO_2^-

to permit the detection of ions outside normal conductivity detector's capabilities or to enhance the specificity of the IC separation. Several good examples demonstrate the enhancement of specificity. Coupled chromatographic methods (described in detail in Chapter 3) increase the specificity of many organic acids because their elution characteristics separated by an ion exchange mechanism are quite different from the profile of the same ions when an exclusion/adsorption/partition separation mechanism is used. Another example of hyphenated ion chromatographic separations is the use of an atomic absorption unit to detect separated arsenic species.[16] Here the inorganic and organoarsenic ions are separated by IC and then detected by an atomic absorption spectrometer (Figure 4.10). One final example of a hyphenated method is the coupling of IC to a reversed phase HPLC instrument.[17] This system has been automated and is used on a process stream that contains 30% total carbon. Sulfate and low levels of chloride are determined.

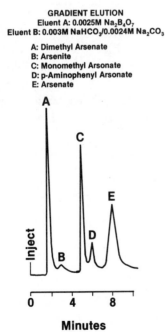

GRADIENT ELUTION
Eluent A: 0.0025M $Na_2B_4O_7$
Eluent B: 0.003M $NaHCO_3$/0.0024M Na_2CO_3

A: Dimethyl Arsenate
B: Arsenite
C: Monomethyl Arsonate
D: p-Aminophenyl Arsonate
E: Arsenate

Figure 4.10. Atomic absorption detection of arsenic ions. Reprinted from *Analytical Chemistry* with permission.[16] Copyright 1981 American Chemical Society.

4.5 OTHER ASPECTS OF NONSUPPRESSED IC

In many cases single-column IC methods offer a good alternative to suppressor-type IC, although some significant column and eluent limitations must be overcome before these methods can become as versatile as suppressor-based IC. Also, single-column methods are not normally quite so sensitive as suppressor-type IC for common anions, although they are sufficiently sensitive for many applications. Employing the same devices that are used in suppressed systems, such as concentrator columns and electrochemical detectors, will enable nonsuppressed IC to expand its sensitivity and analytical range. By applying new detectors and by coupling IC to other analytical methods increased ion specificity can be achieved. In some cases there are unique advantages to hyphenated systems. Use of these methods should increase substantially in the future.

NOTES

1. T. H. Maugh, II, "IC Versatility Promotes Competition," *Science,* **208**(4), 164–165 (1980).
2. R. L. Stevenson and K. Harrison, "Design and Performance of a Modular Chromatograph for Chromatography of Anions," *Amer. Lab.,* May 1981.
3. T. Jupille, M. Gray, B. Block, and M. Gould, "Ion Moderated Partition HPLC," *Amer. Lab.,* August 1981.
4. D. T. Gjerde, J. S. Fritz, and G. Schmuckler, "Anion Chromatography with Low-Conductivity Eluents," *J. Chromatogr.,* **186,** 537–547 (1979).
5. D. T. Gjerde, G. Schmuckler and J. S. Fritz, "Anion Chromatography with Low Conductivity Eluents II," *J. Chromatogr.,* **187,** 35–45 (1980).
6. J. S. Fritz, D. T. Gjerde, and R. M. Becker, "Cation Chromatography Using a Conductivity Detector," *Anal. Chem.,* **52,** 1519 (1980).
7. J. E. Girard and J. A. Glatz, "Ion Chromatography with Conventional HPLC Instrumentation," *Amer. Lab.,* 26, 28, 30, 33, 34, 35, October 1981.

8. D. T. Gjerde, J. S. Fritz, and G. Schmuckler, "Anion Chromatography with Low-Conductivity Eluents," *J. Chromatogr.*, **186**, 537–547 (1979).

9. D. T. Gjerde, J. S. Fritz, and G. Schmuckler, "Anion Chromatography with Low-Conductivity Eluents," *J. Chromatogr.*, **186**, 537–547 (1979).

10. D. T. Gjerde, G. Schmuckler, and J. S. Fritz, "Anion Chromatography with Low Conductivity Eluents II," *J. Chromatogr.*, **187**, 35–45 (1980).

11. J. S. Fritz, D. T. Gjerde, and R. M. Becker, "Cation Chromatography Using a Conductivity Detector," *Anal. Chem.*, **52**, 1519 (1980).

12. Available as a courtesy from Wescan Corp., Santa Clara, California.

13. General Brochure, The Separations Group, Hesperia, California.

14. D. N. Armentrout, J. D. McLean, and M. W. Long, "Trace Determination of Phenolic Compounds in Water by Reversed Phase Liquid Chromatography with Electrochemical Detection Using a Carbon-Polyethylene Tubular Anode," *Anal. Chem.*, **51**(1), 1039 (1979).

15. E. Cathers and E. L. Johnson, "Applications of Simultaneous Detection in IC," presented at the Rocky Mountain Conference, Symposium on Ion Chromatography, Denver, Colorado, August 1981.

16. G. R. Ricci, L. S. Shepard, G. Colovos, and N. E. Hester, "Ion Chromatography with Atomic Absorption Spectrometric Detection for Determination of Organic and Inorganic Arsenic Species," *Anal. Chem.*, **53**(4), 610–613 (1981).

17. T. L. Craven, "Automated Coupled Liquid Chromatography/Ion Chromatography," presented at the Rocky Mountain Conference, Symposium on Ion Chromatography, Denver, Colorado, August 1981.

CHAPTER
5
Ion-Exchange Resins and Ion Chromatography Column Packings

$\mathbf{B}$ecause ion chromatography is a direct offshoot of ion-exchange technology, some knowledge of the many aspects of ion exchangers is helpful. However, an in-depth excursion into ion-exchange theory, thermodynamics, synthesis, and application is quite beyond the scope of this book. Several excellent texts are available in these areas.[1,2] A synopsis of classical ion exchange, especially when pertinent to IC, is offered here. This chapter provides sufficient background to make easier an understanding of the heart of IC: the ion-exchange resins used as column packings. The characteristics of columns in suppressor-type ion chromatography and some of the problems associated with the packings are carefully described. Nonsuppressed IC column packings are also covered. Finally, some future column technology developments are discussed, although this area is somewhat speculative.

5.1 CONVENTIONAL ION-EXCHANGE RESINS

Conventional ion-exchange reactions have been known for many centuries. As early as the writings of Aristotle references are made to certain clays that could "clarify" seawater. Natural ion exchangers such as these zeolite clays, although useful in limited ways, never really found widespread application. It was not until synthetic resins were produced in the twentieth century that ion-exchange technology became commercially important. The first synthesized resins were condensation polymers formed from formaldehyde and various multi-hydroxybenzene compounds. The usefulness of this weak acid ion exchanger led other experimenters to synthesize different resins with increased capacity and better selectivity. Eleven years after the first ion-exchange resin synthesis a resin based on styrene–divinylbenzene (S-DVB) was made. This rugged resin had such versatility that numerous applications were developed.

One very important small-scale scientific application is the column chromatographic separation of ionic mixtures. Ion chromatography is a highly specialized part of this application area. As in large-scale commercial uses, various styrene–divinylbenzene copolymers are presently best types of resin used in suppressor-type IC. Several different kinds of S-DVB resin are commonly used in these column packings. Specifically, IC suppressor columns are packed with conventional gel-type S-DVB resins while specialized pellicular S-DVB packings have been used in separator columns.

Production of gel-type S-DVB resins starts with the S-DVB copolymer, which is made by suspension polymerization of styrene and divinylbenzene. The generalized polymer network is shown in Figure 5.1. The divinylbenzene cross-linking gives the resin its physical strength after chemical capacity has been introduced in subsequent reactions. In the initial polymerization careful control of reaction conditions yields spherical polymer beads in which a large part of the production batch has a reasonably narrow particle-size distribution. Temperature, mixing speed, and many other factors must be carefully controlled and constantly monitored during the process to produce a uniform polymer product. After polymerization the inactive resin beads are rigid hydrophobic spheres that show relatively little difference in physical characteristics based on cross-linking.

The second step in the production of a conventional S-DVB gel-type resin is the chemical reaction of the polymer spheres to introduce ionic

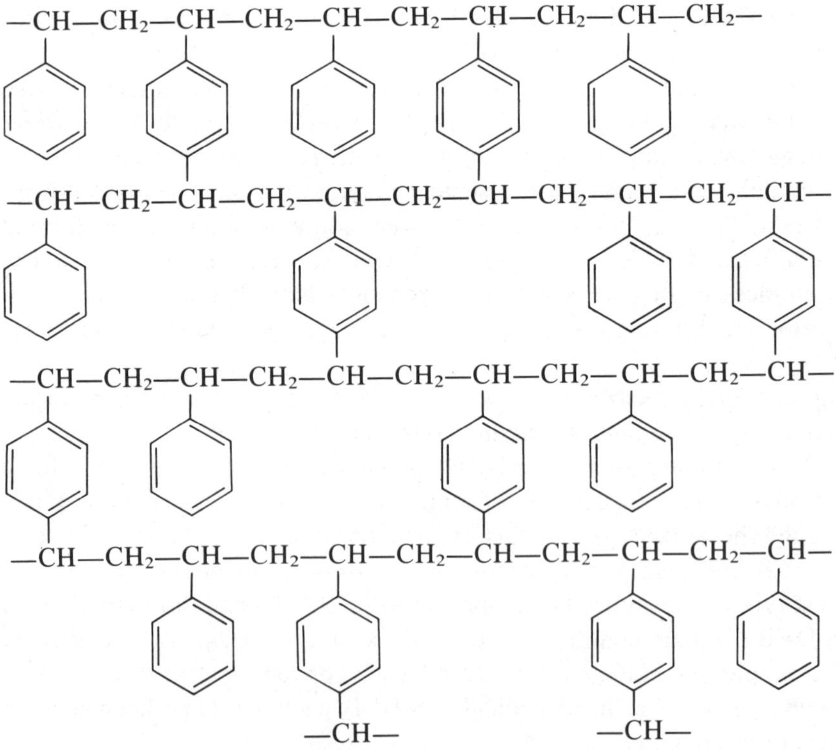

Figure 5.1.   Styrene-divinylbenzene polymer network.

exchange sites. A variety of reactions, each yielding a resin with different chemical properties, has been used to make different common ion-exchange resins. These reactions are summarized in Table 5.1.[3] The two types of resin most commonly used in IC suppressor columns are sulfonic acid (SO_3^-), strong acid resins, and quaternary ammonium $[N^+(R)_3]$, strong base resins.

The sulfonic acid exchange groups are introduced into the conventional S-DVB polymer by a process called sulfonation, which consists of reacting the S-DVB resin beads in hot sulfuric acid, chlorosulfonic acid, or another solvent system that will place the sulfonic acid sites in the aromatic ring of the polymer backbone. The reaction for sulfuric acid is shown in Figure 5.2. The beads are often preswollen with a solvent to decrease process time. Once the ion-exchange resin is formed water is added slowly to swell the beads.

Table 5.1. Chemical Classification of Ion-Exchange Resins

Classification	Active Group
Cation-Exchange Resins	
Strong acid	Sulfonic acid
Weak acid	Carboxylic acid
Weak acid	Phosphonic acid
Anion-Exchange Resins	
Strong base	Quaternary ammonium
Weak base	Secondary amine
Weak base	Tertiary amine (aromatic matrix)
Weak base	Tertiary amine (aliphatic matrix)

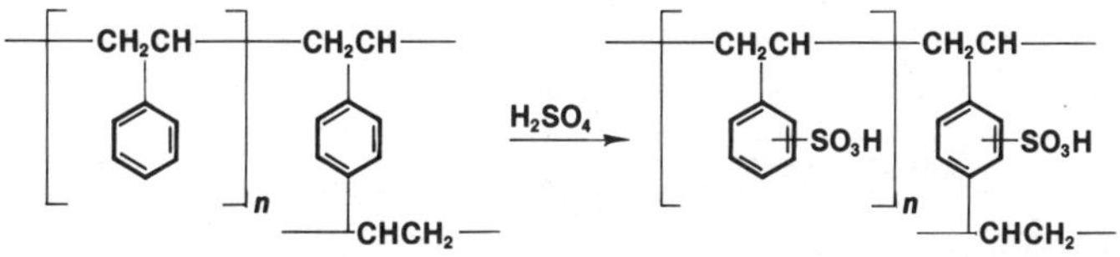

Figure 5.2. Sulfonation of styrene–divinylbenzene. Reprinted from the *Encyclopedia of Polymer Science and Technology* with permission of John Wiley & Sons.

Quaternary ammonium active capacity is introduced into the S-DVB beads in a two-step process. Direct amination of the polymer itself is not feasible; therefore the first step is to modify the resin by forming a chloromethylated intermediate. This intermediate then undergoes aminolysis in a second step with an aminating agent such as trimethylamine or dimethylethanolamine. Figure 5.3 shows the reaction for forming both types of quaternary ammonium ion exchanger. Type 1 is a stronger base than Type 2. There is also the possibility of a complicating side reaction in the formation of the chloromethylated intermediate. This reaction would lead to additional cross-linking in the resin structure; careful control of reaction conditions will minimize this side reaction and yield a relatively uniformly cross-linked resin.

Once the S-DVB resins have been functionalized their physical properties change dramatically. Instead of the small, solid hydrophobic spheres that emerged from the initial polymerization, the beads now have many internal flow paths and will uptake considerable amounts of water or other solvents. This change is caused by the opening up of the resin structure due to electrostatic repulsion of the ionic sites and the requirement that the ionic sites be hydrated. Changes in the polymer chain structure caused by chemical stress also lead to a more open structure.

The amount of water intake is directly related to the divinylbenzene content in the initial copolymer. Resins with low divinylbenzene content (2 to 4% DVB) are not highly cross-linked and will uptake large amounts of water and thus swell to a larger degree than highly cross-linked resins

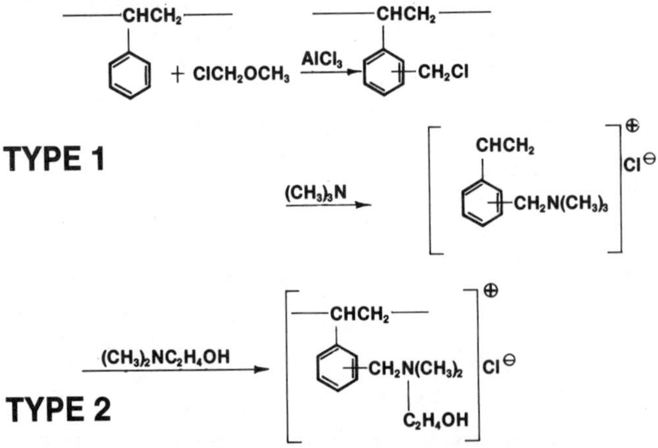

Figure 5.3.   Amination of styrene–divinylbenzene. Reprinted from the *Encyclopedia of Polymer Science and Technology* with permission of John Wiley & Sons.

(10 to 16% DVB). The low cross-linked resins are quite gelatinous and will deform and compress when placed under physical pressure. Resins with 10 to 16% DVB are structurally more rigid and are not easily compressed under pressure. Thus the higher cross-linked resins (8 to 16% DVB) are somewhat more suited for chromatographic column packings.

The introduction of ionic sites into the resin copolymer also drastically changes the chemical properties of conventional gel-type S-DVB resins. The single most important chemical property is capacity. The number of ionic sites per unit weight of resin is termed capacity. This property can be stated for wet or dry weight. Obviously the amount of water that the resin holds will make quite a difference in the values for wet capacity but no difference if the resin has been completely dried. Conventional gel-type resins have an ion-exchange capacity between 4.5 and 5.5 meq/g total on a dry-weight basis. In aminated resins (quaternary ammonium sites) there is a loss with time in the quaternary ammonium capacity from chemical degradation of the resin. Total capacity will remain relatively constant, but there will be an increase in the weak base capacity (from $N - (CH_3)_2$ sites) and a concurrent decrease in the strong base $(N^+(CH_3)_3)$ capacity. Sulfonated resins do not undergo chemical degradation to any significant degree.

A second type of S-DVB resin was also developed several years ago and has recently been applied to IC and ion-exchange chromatographic methods. Macroporous or macroreticular S-DVB resins are made with high (15 to 50%) divinylbenzene content. These beads are extremely rigid and show little deformation under pressure. The pore structures are introduced when the bead is polymerized. Pore-forming agents (poragens), such as isoamyl alcohol/toluene mixtures, are used to create liquid pathways throughout the bead. Careful selection and control of the poragen and polymerization conditions will lead to uniform pore diameters and to the production of uniform packing materials. The nonfunctionalized resins have found considerable use in the removal of organic materials. The resins can be functionalized in the same manner as lower cross-linked gel-type ion-exchange resins.

5.2 ION CHROMATOGRAPHY PACKINGS AND COLUMNS

In the development of IC some confusion has been associated with the use of ion-exchange terminology to describe the resins. First, the IC separator and IC suppressor resins are opposite types of resin (i.e., if one is a cation exchanger, the other is an anion exchanger). This is not the case for ICE resins. Here, both separator resins are cation-exchange resins but in different chemical forms. Second, the suppressor column used in suppressor-type IC is identified *opposite* to the standard identification of the resin. Thus an "anion suppressor" is packed with a cation-exchange resin, whereas a "cation suppressor" is packed with an anion-exchange resin. For the pellicular resins used in IC separator columns the type of resin and the IC system in which it is used are the same; for example, a cation-exchange resin is used in a cation separator.

A great deal of difference exists between IC separator and suppressor resins. Table 5.2 is a comparison of the four types of S-DVB resins used in ion chromatography. The suppressor resins are fully porous gel-type ion exchangers, whereas the separator resins are specialized pellicular ion exchangers.

Conventional gel-type S-DVB resins are used in IC suppressor columns. Resin such as Bio-Rad AG 50W × 12 for an anion suppressor or AG 1 × 10 for a cation suppressor (200 to 400 mesh for both resins) can be purchased from resin manufacturers. The anion suppressor is filled with a cation-exchange resin with SO_3^- functional groups. This resin ex-

Table 5.2.   Comparison of Ion Chromatography Resins

	Resin			
Parameter	Pellicular S-DVB	Macroporous S-DVB	Silica-Based	Gel-Type S-DVB
Use	Anion and cation separation	Anion separation	Anion separation	Exclusion and partition separation of organic species and suppressor columns
Degree of swelling	Small	None	None	Large
Capacity (meq/g)	0.01–0.02	0.007–0.07	0.02–0.1	3–5
Particle size (μm)	15–20	25–35	10–15	37–45 15–20
Useful pH range	0.5–13	0.5–13	2–7	0–14

changes H^+ from the resin for Na^+ in the eluent. The cation suppressor column is filled with anion-exchange resin. This resin exchanges OH^- from the resin for Cl^- or NO_3^- in the eluent. The ICE suppressor and post-suppressor columns use Ag^+ form resin. This resin precipitates $AgCl$ from the eluent while exchanging an H^+ ion for the Ag^+ ion from the resin.

Before being packed into IC or ICE suppressor columns the gel-type resins are narrowed in size distribution by large-scale, water-settling methods. The resins are also further processed to remove contamination. This processing usually consists of a series of acid and base refluxes. Some resin batches also require methanol refluxing. The ICE suppressor resin is converted to the Ag^+ form with $AgNO_3$ and then carefully water-washed before being packed into plastic tubes. After these batch treatments are completed and the columns are packed, the resin again undergoes a series of similar treatments before being tested for chromatographic performance and baseline stability.

In contrast to the IC suppressor resins, the separator resins used for IC analysis are quite unconventional. These resins are the result of long study and extensive development, both in the initial work by Dow Chemical and in subsequent work by Dionex. The exact nature, specific details of preparation, and physical and chemical characteristics of the separator resins are considered proprietary information by Dionex. However, short but adequate descriptions of all separating resins can be made without transgressing into these proprietary areas.

The resins used in IC separator columns must fill two seemingly conflicting purposes: rapid, efficient separation of the ions of interest versus low pressure and long-column lifetimes. To meet these two requirements pellicular resins made from styrene–divinylbenzene have been used as packings.[4,5,6]

A pellicular resin has active capacity only at or near the surface of the bead. This increases efficiency by minimizing the internal volume into which the ions can diffuse. This low-capacity ion-exchange bead has been compared to a billiard ball covered with a thin layer of gelatin. The gelatin layer on the surface of the bead contains all the active capacity. Figure 5.4 shows an expanded view of a sulfonic acid pellicular ion-exchange bead. The two dramatically different regions in the bead create two distinct interfaces for the mobile phase. One interface is between the exterior of the bead and the active capacity in the thin surface shell and the other is between the surface active layer and the rest of the cross section of

CATION SEPARATOR

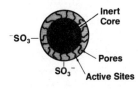

Surface Sulfonated

Figure 5.4. IC separator resin–cation separator. Copyright by Dionex Corp. and reprinted by permission of copyright owner.

the bead. The surface layer of the bead contains all the active capacity and is somewhat similar to conventional gel-type resins. The surface shell (several hundred Angstroms deep) is swollen like the porous resins, thus allowing liquid to flow in and out. The surface polymer structure on the outer surface of the bead forms an electrical film and also acts as a semi-permeable membrane between the inside and outside of the bead for the eluent and sample ions. Because of the small depth of the active layer, eluent flows in and out quite easily. The bulk of the bead is a tight polymeric hydrocarbon structure and as such is hydrophobic and does not swell in water-based eluents. This minimizes diffusion of the sample ions and eluent ions into the bulk of the beads. It also prevents the separator resins from swelling excessively in organic solvents such as methanol or acetonitrile. Because pellicular resins are highly efficient in separating ions, relatively large beads (15 to 20 μm) can be used. This helps to keep system operating pressures low (typically less than 750 psig).

The surface-sulfonated S-DVB materials are used as the cation-separator resin and substrate for the two-part anion-separator resin. An aminated latex completes the anion-separator resin.

The low capacity of the cation-separator resin allows use of low-strength ionic eluents, which enhances IC detection of trace amounts of cations and is absolutely necessary for cation-exchange separation methods that do not include a suppressor column. The pellicular nature of the resin permits rapid diffusion of the ions into and out of the active surface of the bead. The short diffusion pathways result in excellent mass transfer characteristics in the resin and yield good separations for a variety of cations. Depending on the eluent, the cation-separator resin can be used to separate group I metals, ammonia, small aliphatic amines, and quaternary ammonium compounds or group II metals and some large amines.

The packing in the anion separator column used in suppressor-type IC is made up of two types of resin combined into one. The surface-sulfonated S-DVB resin used as the cation-separator packing is the substrate for this two-part resin. An aminated latex is then coated onto the surface of the substrate to create the anion-separator resin.

The need for a high efficiency aminated resin was made quite clear during the initial development of the ion chromatograph. Unlike sulfonation, the process of chloromethylating the backbone polymer and then converting it to a quaternary ammonium ionic site proved to be difficult to control. A pellicular aminated resin was made by attaching a micro-particle aminated latex to the surface-sulfonated resin. This yielded a highly efficient anion-separator packing in a patented process that has been termed agglomeration.[7,8]

Figure 5.5 is a representation of this resin.[9] A scanning electron microphotograph is shown in Figure 5.6. The latex is held on the surface primarily by the electrostatic attraction of the positively charged latex particles to the negatively charged surface active sites and by Van der Wals forces, although other forces may affect the bonding. Once again excellent mass transfer and low internal and interstitial volume result in sharp peaks and good separation. The resulting resin has low capacity and high selectivity for ions which allow low eluent concentrations to be used in the separation of anions. There appears to be a need to control

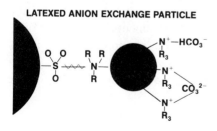

Figure 5.5. IC separator resin-anion separator. Copyright by Dionex Corp. and reprinted by permission of copyright owner.

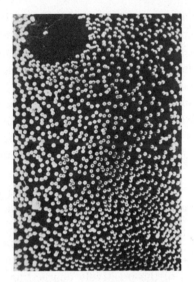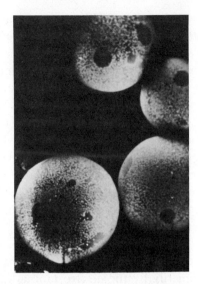

Figure 5.6.   Microphotograph of anion separator resin. Courtesy of Dow Corp.

the latex volume to substrate volume ratio which prevents large increases in separator capacity as the substrate size is lowered from 25 to 15 μm in size. At the 25 μm size a 0.1 to 0.2 μm latex particle is used. At the 15 μm size a 0.02 μm latex particle is used. This arrangement yields a highly efficient column[10]. The resin also has some residual cation exchange capacity from SO_3^- sites within the active shell of the substrate. It has been reported that anion-separator columns that no longer give adequate anion separation may be used for cation separation.[11]

In contrast to the packings in IC separators, the ICE separator columns are packed with conventional, moderate cross-linked cation-exchange S/DVB resin. Low flow rates are necessary because the early column packings were low cross-linked resins, which are compressible. New ICE separator columns use much smaller particles with moderate cross-linkage. However, low flow rates must still be used to keep the columns within the pressure limits of the suppressor-type instruments. If the flow rates are excessive, system pressure will rise until a component failure is experienced. This is most often a liquid line or the column itself. These columns should be carefully monitored for headspace development or for flow-rate changes.

5.3 COLUMN PROBLEMS AND TREATMENTS

A number of technical reports are available from the Dionex Corporation on columns and their problems and rejuvenation/storage procedures.[13-16] In general, the resins in IC suppressor columns are quite trouble-free. These columns will last through many thousands of injections before replacement is required. Although they are quite rugged some problems have been observed, the most common of which is baseline noise at high detector sensitivity (low μMHO setting). Styrene–divinylbenzene resins have long been known for "color throw." If these resins are allowed to sit in contact with water or other solvents, the liquid layer often becomes colored. Several explanations have been offered to explain this phenomenon. Contaminants in the starting materials may leach out or small, linear unpolymerized alkylsulfonates may work their way out of the matrix. Another possibility to explain color throw is that parts of the resin structure are broken off by oxidation or photolysis. This "column noise" is minimized by the treatment given to the columns after packing and by proper treatment when the columns are stored. Column noise should not be confused with the electronic noise that is characteristic of the instrument's conductivity detector. If column noise occurs in any IC suppressor, the best treatment is to cycle the suppressor several times through four separate treatments: strong eluent, water, regenerant, and water wash. Degassed water may help to minimize gas-bubble formation and its related baseline noise (the detector reacting to each pump stroke).

Three other baseline stability problems are possible. The first is the presence of gas in the column. Entrapped gas in the suppressor causes baseline pulsations. The compression and expansion of entrapped bubbles allow them to move through the column and eventually be eliminated. Introduction of a large amount of air into the column may also cause channels to form in the resin bed. These channels are preferential flow paths that can produce apparent loss in capacity (shorter than usual interval between regeneration cycles). Channeling requires long pumping times to restore the resin bed operation. Strongly retained species are a second source of column baseline noise. Examples of these types of ion are ClO_4^- for cation suppressors and Ba^{2+} for anion suppressors. The slow elution of these species will often result in domelike disturbances in the baseline in later injected samples. Even with counterflow regeneration these species may be difficult to remove without long regeneration times.

Strongly oxidizing species such as ClO_4^- may also result in oxidation of the resin itself. Normal regeneration for IC suppressors is set up with flow countercurrent to the eluent flow path. This will minimize this type of baseline noise because the strongly retained species are relatively easily removed from the short length of resin at the top of the suppressor.

The last type of column-related problem results from reactions that occur between sample ions and the eluent or regenerant ions or from the interaction of sample ions with the ionic sites or polymeric structure of the suppressor column. An example is the precipitation of transition metals and other species as hydroxides in the cation suppressor. One notable exception is the elution of Ni^{2+} ion by standard divalent cation conditions.[17] Another example is the surface fouling of the resin beads with particulate materials or precipitates. These problems are often difficult to identify but can sometimes be minimized by chemical treatments. Acid-base cycling of the resin and prolonged water wash are two possible treatments.

In the ICE system regeneration and resilvering of the suppressor resin is quite lengthy and not easily done with the conventional IC regeneration system. For this reason ICE suppressors are made to be disposable. Another reason is that the precipitated AgCl causes pressure buildup in the column, which can be eliminated if the expended portion of the suppressor is cut off after the column is allowed to sit without eluent flow and before its next use. The post-suppressor column is filled half with silver form resin (top) and half with H^+ form resin (bottom). The suppressor column should be changed before the HCl breaks through from the suppressor onto the post-suppressor. Otherwise, substantial pressure increases will be observed.

Because pellicular column packings are somewhat fragile, precolumns are used in series in front of the separator column. These short precolumns are packed with materials similar to those in the separator columns and serve to protect the separator column by trapping particulate materials from the samples, sorbing organic materials, and retaining polyvalent ions that would decrease the capacity of the separator column.

The precolumns should be periodically cleaned to prevent overloading with contaminants. The frequency of cleaning will depend on the nature of the samples run and the care of the operator. Several routine treatments have been developed. The anion precolumn should be washed with $0.1M$ Na_2CO_3 every two to three weeks, depending on use. The cation precolumn should be treated with $1M$ HNO_3 at the same interval. If either type of column does not perform acceptably after these treatments, an overnight soak in $2M$ H_2SO_4 for anions and $0.5M$ NaOH for cations should be provided. These treatments should be followed by a 30-min water wash, a 15-min wash with strong electrolyte ($0.1M$ Na_2CO_3 for anions and $1M$ HNO_3 for cations), another 15-min water wash, and finally by equilibration with the appropriate eluent.

If these procedures are not satisfactory, washing with $0.01M$ sodium citrate or $0.01M$ sodium tartrate is an alternative. One last-ditch effort can be made if all else fails. This is a 5-min wash with $1M$ HCl/50% methanol or acetone, which should be collected in a beaker. This treatment should not have to be repeated more than twice a month.

Several problems that affect IC separator column performance have been observed. They are associated with sample-eluent chemical interactions or with physical reactions caused by the sample. Use of precolumns should help to minimize them.

One of the most common is contamination from sample ions. A high concentration of transition metals or multivalent cations will produce severe capacity loss in the cation separator. These ions can be removed by flushing with $1M$ HNO_3 for 30 min and another 30-min water wash. Another source of cations is the eluent itself (even if ultrapure HNO_3 or HCl is used) or ions leached from the pump head or gauge. One method of minimizing this difficulty is to use a trapping column in the eluent stream before the injection valve. These scavenger columns must also be cleaned periodically to remove accumulated contaminants.

Conditions similar to the loss of capacity in the cation system also develop in the anion separators. Anion separator resins can be affected by loading with strongly retained multivalent species such as CrO_4^{2-} and ClO_4^-. These species are slowly eluted with normal eluents. Species that precipitate in the high pH eluents also cause trouble, as do chemical interactions of the column and sample. Cation species are exchanged onto the unused cation capacity of the IC anion-separator column. These ions can subsequently affect separations. One example is the presence of Fe^{+3} and Cu^{+2} ions in the sample. These ions are reported to affect SO_3^{2-} analysis by oxidizing the SO_3^{2-} to SO_4^{2-} on the column.[18] Overloading the column with sample ions or with organic solvents is also common. Because the separator capacities are low, even moderate (about 1000 to 2000 ppm) concentrations of species such as Cl^- or SO_4^{2-} ions begin to load up the column. High concentration of ions can also act as eluents. The result is lower retention of ions that have lower selectivity than the major sample ion. This can cause the less well retained species to elute closer to the void volume. If ion concentrations for major species exceed 0.5% (5000 ppm), the sample should be diluted before any IC injection to prevent overloading.[19] Organic solvents such as alcohols, ketones, or aromatic-aliphatic hydrocarbons will often cause the resin beads to swell or develop channels. Sometimes these physical reactions are irreversible. The best way to handle samples that contain these materials is to dilute or extract the sample. If this is not practical, certain types of alcohol and ketone can be added stepwise to the eluent. The concentration of the organic solvent should be gradually increased to a maximum of 45 to 50% by volume, although uses as high as 100% have been reported.[20] Higher concentrations of the organic solvent may cause unpredictable, often irreversible changes in the separator column. Return to an all-water-based eluent should also be stepwise with adequate equilibration for each step.

5.4 NONSUPPRESSED ION CHROMATOGRAPHY MATERIALS

Nonsuppressed ion chromatography has been developed to complement suppressor-type IC. These methods were introduced to eliminate the need for the suppressor column and have both technical and commercial promise. Although nonsuppressed IC is generally not as sensitive as suppressor-type IC, it is capable of many ion analyses.

Nonsuppressed IC separator packings are now two distinct types. The first packing material is a silica-based ion-exchange resin. This pellicular packing is marketed under the tradename Vydac® by The Separations Group (Hesperia, California) and can be used with acidic or nearly neutral eluents in conventional HPLC systems with auxiliary conductivity detection. Column capacities are about 0.1 meq/g. These columns must be treated with some care to avoid damage from overloading or to prevent damage to the resin itself from the sample or the eluent. High pH samples or high salt content may irreversibly damage the column packing. Column treatments consist of washing with higher concentrations of eluents or organic solvents like DMSO or acetonitrile. The second type of separator column has been used in conjunction with a conventional HPLC and an auxiliary conductivity detector. This column is packed with macroporous S-DVB resin specially synthesized for anion separations.[21] The surfaces of the internal pores are chloromethylated and then aminated in the same manner as gel-type resins. Exchange capacities are low, and the column is quite useful with potassium benzoate and potassium acid phthalate eluents (the same eluents used with silica columns). This packing should be as rugged as the separator resins used in suppressor-type columns. It was not yet commercially available when this text was prepared.

5.5 FUTURE DEVELOPMENTS IN COLUMN PACKINGS

The variety of column packing materials designed for use in IC should increase in the future. Silica and macroporous S-DVB resins have different affinities for various ions, but both appear to have similar affinities for common ions. The biggest changes should occur in resin selectivities which will involve the functionality of the resin or clever innovations, such as mobile phase ion chromatography.[22] Resins specifically synthesized to solve certain separation problems (such as bromide ion in high chloride brines with an SA-2 separator) or optimized for certain applications will undoubtedly be introduced by resin and instrument manufacturers. Higher efficiency columns will yield improved separations. Improved column treatments and rejuvenation procedures will extend the lifetimes of separator columns. Disposable, lower cost columns will allow IC users to run more complex, higher concentration samples. On-column sample pretreatments will also permit the injection of more concentrated samples. Novel uses of detectors other than suppressor column/conductivity and of many different eluent-resin combinations should greatly increase the range of IC separations.

NOTES

1. W. Reiman III and H. F. Walton, *Ion Exchange in Analytical Chemistry*, Pergamon, Oxford, 1970.
2. F. Helfferich, *Ion Exchange*, McGraw-Hill, New York, 1962.
3. I. M. Abrams and L. Benezra, "Ion Exchange Polymers," *Encyclopedia of Polymer Science and Technology*, Wiley, New York, 1967.
4. C. A. Pohl, and E. L. Johnson, "Ion Chromatography-State of the Art," *J. Chromatogr. Sci.*, **18**(9), 442–452 (1980).
5. P. Hajos and J. Inczedy, "Preparation and Ion Chromatographic Application of Surface-Sulfonated Cation Exchangers," *J. Chromatogr. Sci.*, **80**(201), 253–257 (1980).

6. T. S. Stevens and H. Small, "Surface Sulfonated Styrene Divinyl-benzene—Optimization of Performance in Ion Chromatography," *J. Liq. Chromatogr.*, **1**(2), 123–132 (1978).

7. H. Small and T. S. Stevens, U.S. Patent No. 4,101,460.

8. R. Chang and F. Smith, U.S. Patent No. 4,119,580.

9. Dionex Training Course (1978).

10. C. A. Pohl and A. Woodruff, "High Performance Ion Chromatography (HPIC)," presented at the Rocky Mountain Conference, Symposium on Ion Chromatography, Denver, Colorado, August 1981.

11. J. Wimberly, "Ion Chromatographic Separations of Cations in an Anion Separator Column," *Anal. Chem.*, **53**(11), 1709–1710 (1981).

12. "Ion Chromatography Separator Column Packings," Dionex Corporation, Sunnyvale, California, IC Exchange 3, August 1980.

13. "Long Term Storage of IC Columns," Dionex Corporation, Sunnyvale, California, Technical Note 7.

14. "Routine Maintenance and Reconditioning of Columns," Dionex Corporation, Sunnyvale, California, Technical Notes 2 and 2R.

15. "Use of Small Suppressor Columns," Dionex Corporation, Sunnyvale, California, IC Exchange 2, March 1980.

16. "Why Use A Suppressor Column?" Dionex Corporation, Sunnyvale, California, IC Exchange 4, August 1980.

17. A. Fitchett, Dionex Corporation, Sunnyvale, California, private communication.

18. L. D. Hansen, B. E. Richter, D. K. Rollins, J. D. Lamb, and D. J. Eatough, "Determination of Arsenic and Sulfur Species in Environmental Samples by Ion Chromatography," *Anal. Chem.*, **51**(6), 633–637 (1979).

19. M. A. O. Bynum, S. U. Tyree, Jr., and W. E. Weiser, "Effects of Major Ions on the Determination of Trace Ions by Ion Chromatography," *Anal. Chem.*, **53**, 1935–1936 (October 1981).

20. E. L. Johnson, "Care and Feeding of IC Columns," presented at the Rocky Mountain Conference, Symposium on Ion Chromatography, Denver, Colorado, August 1981.

21. D. T. Gjerde, J. S. Fritz, and G. Schmuckler, "Anion Chromatography, Using Low Conductivity Eluents," *J. Chromatogr.*, **186**, 537–547 (1979).

22. C. A. Pohl and M. Ebenhahn, "Mobile Phase Ion Chromatography (MPIC)," presented at the Rocky Mountain Conference, Symposium on Ion Chromatography, Denver, Colorado, August 1981.

CHAPTER
6
Applications of Ion Chromatography

Ion chromatography has been successfully applied to the analysis of ions in many extremely diverse types of sample. Determination of ions in difficult sample matrices such as toothpastes, brines and caustics, and Kraft black liquors has become common IC practice. Ions in simpler matrices such as rain and potable waters are also directly analyzable by IC. Although IC is used primarily in the analysis of ions at 0.5 to 100 mg/L, it has also been extended to analyses below 10 μg/L in such samples as ambient SO_2 gas and ultrapure steam.

This chapter covers IC work from the viewpoint of areas of application. It is not meant to be a rigorous treatment of IC applications. Numerous review papers have been written on IC.[1-23] In some cases improvements in methods may exist that have not yet been brought into common practice. Also, the applications of IC are expanding and changing so rapidly that associating a specific analytical procedure with an application is impractical. In some cases only the feasibility of performing certain separations or applications has been demonstrated. Problems with interferences and column degradation have not been resolved to the point that the method can be considered proven. It is hoped that the versatility of ion chromatography as an analytical technique is demonstrated in this chapter.

Since the initial publication of the feasibility of ion chromatography the number of ions that can be determined by IC has increased dramatically. Tables 6.1 and 6.2 are summaries of the ions that are *presently* determinable by ion chromatographic methods. Inorganic anions have shown the greatest increase in numbers because most application efforts have been directed toward these species. Development of IC methods for cation determination have lagged behind considerably but are presently moving foreward rapidly.

For all the breadth of types of ion and for the great range of ion concentrations that can be determined by IC the ion chromatograph is used primarily as a 0.01 to 50 mg/L sulfate and chloride analyzer. Because so many applications involve the determination of sulfate ion in different matrices, this ion naturally occurs on most of the typical chromatograms that illustrate the various applications. It is also the basic nature of most ion chromatograms to be relatively simple, often with only seven or fewer peaks. Thus many of the chromatograms used in this chapter to illustrate specific applications will look much like the chromatograms for synthetic

Table 6.1. Types of Anion Presently Determinable By IC

Halides	Phenolics
Oxyhalides	Phosphorous oxides
Sulfur oxides	Metal oxides
Nitrogen oxides	Boron compounds
Organic acids	Chlorinated phenolics

Table 6.2. Types of Cation Presently Determinable by IC

Alkali metals
Alkaline earth metals
Aliphatic amines
Transition metals

standards. Also, most IC work done to date represents in large part the use of standard or nearly standard instrument-operating conditions. These conditions are not noted on the various chromatograms except when significant changes were made from standard conditions.

6.1 AIR POLLUTION

The first area in which IC found widespread acceptance was in air pollution analysis. In this area analytical methods had not become so deeply entrenched as in other phases of analytical chemistry. Because the ion chromatograph solved a variety of analytical problems and because of its versatility, IC was soon in general use in many air pollution analyses beyond the initial single application.

The first type of air pollution samples to be analyzed by IC were atmospheric particulates trapped on high volume filters.[24-44] Sections of these filters were extracted in water, IC eluents, or other buffers. Most extractions used ultrasonic agitation. The quantitative nature of the extractions was verified by repetitive extraction. Fluoride, chloride, nitrate, and sulfate were detected on the first filters (Figure 6.1).[45] Among these ions nitrate and sulfate were of the greatest interest. Comparison studies between the ion concentrations determined by IC and those obtained by the cadmium reduction method for nitrate ion and the methyl thymol blue

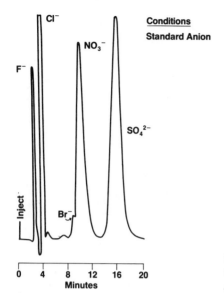

Figure 6.1. Ambient particulate analysis for anions. Reprinted with permission of *Ann Arbor Science.*[47]

method for sulfate ion showed reasonably close agreement.[46] Discrepancies in the NO_3^- data have been explained by the presence of NO_2^- ion in the samples. This ion is not distinguishable from NO_3^- by the wet chemical method but is quite widely separated from NO_3^- by IC. Low levels of NO_2^-, however, are not easily detected by IC when standard anion instrument conditions are used. No nitrite IC determinations were done during the IC analyses of these samples.

The next area of air pollution analysis to which IC was applied was aerosol characterization.[47,48,49,50] Aerosols are quite closely related to ambient air particulates in that they consist in part of small-diameter particles that remain suspended in the atmosphere in gas or smoke clouds or fog/smog. In addition to the particles, aerosols have entrapped or entrained the gases or liquids that surround the particles. Of special interest are the H_2SO_4 aerosols, which are damaging to plants and human and animal health and can cause environmental damage such as metal corrosion or plastic and rubber cracking. Benzaldelyde and several other reagents have been used in attempts to isolate the H_2SO_4 aerosols from other sulfate species [Na_2SO_4 or $(NH_4)_2SO_4$] which may be present in the filters. Some success has been reported in the isolation of acidic aerosol with perimidylammonium bromide as a selective extractant.

Soon after IC was applied to particulate and aerosol analyses another important air pollution use was developed. The multi-ion separation ca-

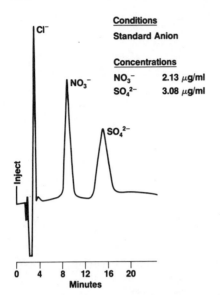

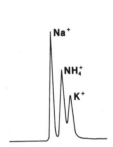

Figure 6.2. Acid resin analysis. Reprinted with permission of *Ann Arbor Science*.[45]

Figure 6.3. Cation content of acid rain. Reprinted with permission of *Ann Arbor Science*.[55]

pabilities and the ability of IC to detect extremely low ion concentrations made acid rain analysis a natural application.[51–59] Once again, the initial IC separations were done under standard instrument conditions. Both anions (Figure 6.2) and cations (Figure 6.3) were analyzed routinely by IC. These instrument conditions were later modified by the use of lower volume suppressor columns and higher eluent flow rates, by eliminating liquid dead volumes, and by using a pulseless eluent pump.[60] Concentrator columns have been used to lower the minimum detection limits when large (10 to 25 mL) sample volumes were available. These columns have also been proposed for automatic on-site loading of rainwater samples in a precipitation event. The ion chromatograph has been so widely used in acid rain analyses that it has been named as the method of choice.[61]

In addition to the ions normally found in air samples, higher concentrations or different ion profiles may result from localized conditions. In one example high fluoride concentrations were observed in rain that fell in the vicinity of a chemical plant after an accidental HF gas discharge accident. The fluoride ion is easily detected in standard anion IC analyses.[62]

Besides acid rain, ion chromatography has been applied to the analysis of gases that form acid rain anions. One of these gases is sulfur dioxide

(SO_2).[63-68] Ambient atmospheric levels of SO_2 gas vary widely in concentration. For IC analyses samples of gas are passed through an impinger train with bubblers that contain 3% H_2O_2/0.003M $NaHCO_3$/0.0024M Na_2CO_3. The SO_2 gas is quantitatively converted to SO_4^{2-} ion, which is easily determined by IC (Figure 6.4).[69] Concentrator columns have been used to extend the range of IC to low levels of SO_2 gas. The IC method agrees well with SO_2 concentrations obtained by the modified West-Gaeke method. Studies have shown that the IC scubber solutions are stable for at least one week and that they do not exhibit the temperature dependence of the West-Gaeke method.[70] Also, the IC scrubber does not use the tetrachlormercurate ion.[71]

IC analysis of nitrogen oxide gases has not been so successful as SO_2 gas analysis. Apparently it is more difficult to convert NO and NO_2 gases to ionic form. Nitrogen values by IC are quite a bit lower than by conventional wet chemical methods.[72] However, in one paper samples were successfully analyzed by IC for NO_2.[73]

One major IC application in air pollution is the separation of sulfur oxide anions.[74-82] This analysis is crucial in preventing air-quality deterioration from sulfur oxide gases emitted by coal-fired power plants. Flue-gas desulfurization (FGD) processes have been developed to remove sulfur oxide gases. Two main gases are of interest in the stack emissions: SO_2 and SO_3, indicated by the presence of SO_3^{2-} and SO_4^{2-} ions in the

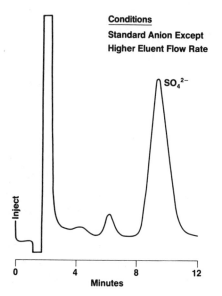

Figure 6.4. Ambient levels of SO_2 gas as sulfate ion. Reprinted with permission of *Ann Arbor Science*.[69]

Conditions

Standard Anion

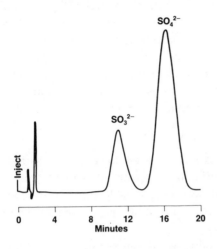

Figure 6.5. Sulfite–sulfate separation. Copyright by Dionex Corp. and reprinted by permission of copyright owner.

Table 6.3.   Recoveries of NO₂ and SO₂

Amount of sample collected	Recovery (%)		Relative Standard Deviation (%)	
	NO₂	SO₂	NO₂	SO₂
31.1 μg of NO₂	80.1	95.8	+ 13.1	
13.5 μg of SO₂				+ 14.2
62.2 μg of NO₂	102.7	99.7	+ 11.7	
27.0 μg of SO₂				+ 8.0
93.3 μg of NO₂	96.3	97.5	+ 4.8	
54.0 μg of SO₂				+ 8.7
124.4 μg of NO₂	106.5	100.3	+ 2.4	+ 3.4

Reference 83. Reprinted with permission of *Anal. Chem.*

IC analyses of trapped samples. Before IC analysis the gases are trapped in an impinger train and converted to ionic form. The ion chromatograph reveals a simple straightforward separation for SO_3^{2-} and SO_4^{2-} ions (Figure 6.5). Table 6.3 is a summary of recoveries for SO_2 and NO_2 gases.[83]

Some complications in the analysis of the SO_3^{2-} ion in solution have been reported. Sulfite solutions are quite unstable and must be prepared frequently. Addition of formaldehyde or mineral acids to SO_3^{2-} solutions helps to reduce oxidation of SO_3^{2-} to SO_4^{2-} [84,85] The presence of certain multivalent cations in the samples also has been reported to cause more rapid oxidation of SO_3^{2-} ion.[86] These cations are exchanged onto the anion separator substrate's sulfonic acid active sites and apparently catalyze SO_3^{2-} oxidation. Another problem reported in this paper is the oxidation of the SO_3^{2-} ion caused by oxygen which is present in the instrument. It is postulated that the Teflon tubing used in suppressor-type ion chromatographs is sufficiently permeable to O_2 gas to allow enough of this oxidant to come into contact with the SO_3^{2-} ion to convert some of the SO_3^{2-} to SO_4^{2-}.

The ion chromatograph has been used in several other ion analyses connected with FGD scrubber operations. In the lime/limestone scrubbers, calcium and magnesium ions in the fluid bed are analyzed by divalent cation IC analysis.[87] This analysis is critical to bed operation because the balance between these ions dictates which sulfite or sulfate salts are formed. If these cations become depleted or if there is a large shift in the cations present, the scrubber bed will lose scrubbing efficiency. Additionally, a loss in the fluidity of the bed may occur, which can cause extremely severe operating problems. The calcium and magnesium salts that are formed in the bed can quickly set up like cement. This failure could result in long downtimes and expensive facility repairs. Figure 6.6 shows the divalent cation analysis of these ions in a diluted FGD stream sample.[88]

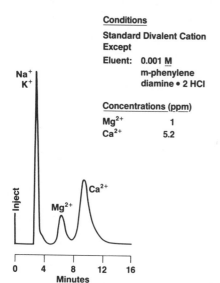

Conditions

Standard Divalent Cation Except

Eluent: 0.001 M m-phenylene diamine • 2 HCl

Concentrations (ppm)

Mg^{2+}	1
Ca^{2+}	5.2

Figure 6.6. Calcium and magnesium ions in limestone scrubber feed. Copyright by Dionex Corp. and reprinted by permission of copyright owner.

Another type of scrubber is used at some power plants to remove sulfur oxide gases. The dual alkali scrubber system uses different contact materials to convert the gases to ionic form. A mixture of NaOH and Na_2CO_3 traps and converts the SO_2 and SO_3 gases. The resulting solution is then pumped to another part of the scrubber system where calcium and magnesium ions precipitate the sulfur oxide ions. Once again, SO_3^{2-}, SO_4^{2-}, CO_3^{2-}, Ca^{2+}, and Mg^{2+} ions are routinely monitored by IC.[89] In addition to these ions, adipic acid and its degradation products can be monitored by ICE analysis. The adipic acid is added to enhance sulfur oxide gas removal. Figure 6.7 shows that adipate ion can be separated from SO_3^{2-} and SO_4^{2-} in a standard IC analysis.[90] However, adipate degradation products (possibly glutarate ion) pose potential interference problems for SO_3^{2-} ion analysis by anion IC. Anion ICE has been reported to be effective in separating these organic acids.[91]

Ion chromatography has also been applied to ion determinations from mobile emission sources which include automobiles, diesel engines, aircraft, and ship exhausts.[92-97] As in other air pollution applications, gases are trapped in impinger trains and particulates are removed by filters. In one special case the IC sulfate analysis method has been used to analyze gases and particles trapped by a sampler specially designed for roadside collection of samples.[98] Sulfate ion is the most common ionic component in these samples.

A variety of air pollution-related analyses have been developed with the ion chromatograph. Particles collected at high altitudes[99] and near the earth's surface level[100] have been analyzed by IC for anion content. Atmospheric volcanic dust from Mt. St. Helens has been analyzed for a

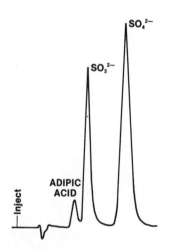

Figure 6.7. Adipic acid–sulfite–sulfate separation.

variety of anions and cations.[101] Fluoride gases and particulates collected near brickworks and aluminum smelters[102] and anions in air particle samples from copper smelter areas[103] and nickel smelters[104] have been routinely analyzed by ion chromatography. Sulfur species extracted from coal fly ash,[105] potential groundwater contamination by FGD wastes,[107] and ammonium sulfamate[107] and azide[108] in air samples have been run on an ion chromatograph. Finally, IC results for sulfur and arsenic species are compared with proton-induced X-ray emission analyses and the discrepancies in data have been debated.[109,110]

Many of the details of these air pollution analyses are covered in the papers presented at the two Environmental Protection Agency conferences on Ion Chromatographic Analysis of Environmental Pollutants. The proceedings of these meetings are published in two books. Volume I is almost entirely devoted to air pollution analyses.[111] Included are papers on the comparison of IC analysis to other analytical methods, IC sensitivity and reproducibility, IC analyses at ultratrace levels from filter blanks, and anion separator column lifetimes and problems. This conference was held when IC was still a new method. Volume II contains papers presented a year later at a second symposium.[112] There are still quite a few papers on air pollution applications of IC in this second book. Other papers cover industrial hygiene analyses, water quality applications, industrial use of IC, soil analysis, power-plant applications, and research to extend the IC method.

Research by EPA or under EPA grants and contracts aimed at extending the air pollution applications of IC analysis is continuing at a slow pace. Ion chromatographs are now widely used in acid rain, ambient particulate, and aerosol analyses and have been designated as the method of choice for level 1 screening assessment measurement of anions and some cations. In most air-pollution-related applications of IC standard or nearly standard instrument-operating conditions are sufficient to produce adequate separation of the ions of interest.

6.2 INDUSTRIAL HYGIENE, TOXICOLOGY, AND FORENSIC SCIENCE

Many IC applications related to industrial hygiene/toxicology involve the analysis of ambient gases. The resulting samples and associated IC methods are quite similar to those of many air pollution IC methods. Once again anionic species are the most frequently analyzed by IC.

Arsenate ion (AsO_4^{3-}) has been observed in ion chromatograms for samples collected near copper smelters.[113-116] This ion apparently has significantly different elution characteristics, depending on the columns and eluents used. In the three IC analyses reported the AsO_4^{3-} ion elutes well before SO_4^{2-} or somewhat after SO_4^{2-}. Figure 6.8 shows one reported elution pattern for arsenate ion.

Other arsenic species have been separated by IC and detected by a method different from conductivity. One analysis deals with the determination of organic and inorganic arsenic ions.[117] These species can be separated by IC columns and eluents, but their acid forms have insufficient conductivity to be detectable by the conductivity cell in the ion chromatograph. An atomic absorption spectrophotometer is used as the detector. A continuous gradient with $0.0025M$ $Na_2B_4O_7$ and $0.003M$ $NaHCO_3/0.0024M$ Na_2CO_3 produced the best separation of the five arsenic species. The effluent from the suppressor column is pumped into the spectrophotometer, where an arsine generator converts the arsenic ions to arsine for detection. Proton-induced X-ray emission analysis has also been used for arsenic and sulfur analysis in the same type of samples. Comparison to IC analysis showed some substantial variations. Losses in sulfur by the X-ray emission analysis method has been reported and debated.[118,119]

A second industrial hygiene/toxicology application monitors industrial atmospheres for contaminants. One important use of IC is the determination of formaldehyde and formic acid vapors.[120-124] The formaldehyde–

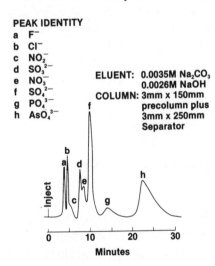

PEAK IDENTITY

a F^-
b Cl^-
c NO_2^-
d SO_3^{2-}
e NO_3^-
f SO_4^{2-}
g PO_4^{3-}
h AsO_4^{3-}

ELUENT: 0.0035M Na_2CO_3
0.0026M NaOH
COLUMN: 3mm x 150mm
precolumn plus
3mm x 250mm
Separator

Figure 6.8. Separation of arsenate. Reprinted from *Analytical Chemistry* with permission.[116] Copyright 1979 American Chemical Society.

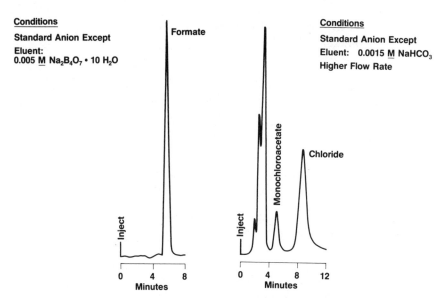

Figure 6.9.   Formaldehyde analysis with IC. Copyright by Dionex Corp. and reprinted by permission of copyright owner.

Figure 6.10.   Chloride and monochloroacetate in industrial atmospheres. Reprinted from *Analytical Chemistry* with permission.[129] Copyright 1979 American Chemical Society.

formic acid is trapped in personnel sampling tubes. These species are subsequently extracted from the packing and analyzed by IC. The formate ion can be eluted with 0.0025 or $0.005M$ $Na_2B_4O_7$ eluents (Figure 6.9).[125] Several varieties of trapping media have been evaluated.[126,127] Once again the ion chromatograph has become the recommended analytical method for these and several other species. Along this line hydrogen cyanide has been determined after it is converted to formate ion.[128]

A method similar to that for formaldehyde has been developed to measure chloroacetylchloride (CAC) in polymer processing plants.[129] CAC is trapped in tubes loaded with silica gel. After sampling the packing is removed from the sampling tube to a $0.0015M$ $NaHCO_3$ solution. After 30 min of ultrasonic agitation the extracted sample is allowed to sit for 4 h to complete conversion of CAC to chloroacetate and chloride ions. The sample is then injected into an ion chromatograph with $0.0015M$ $NaHCO_3$ as the eluent. Figure 6.10 is a typical ion chromatogram of this analysis.[130] The chloroacetate and chloride ions are widely separated and their quantitative values are directly related to CAC concentration, as shown in Table 6.4.

Table 6.4.　Results for Validation of CAC Method

CAC Air Concentration (ppm)[a]	Expected Chloroacetate Concentration[b] (μg/mL)	Found Chloroacetate Concentration[c] (μg/mL)
0.83	31.6	30.0 (0.013)
0.77	29.5	29.2 (0.029)
0.42	15.8	16.1 (0.016)
0.39	14.8	15.2 (0.033)
0.17	6.3	6.3 (0.063)
0.15	5.9	6.4 (0.039)
0.083	3.2	3.1 (0.033)
0.077	3.0	3.1 (0.032)
0.042	1.6	1.6 (0.156)
0.039	1.5	1.5 (0.133)

[a] Based on a 20-L air sample.
[b] Based on weight losses from CAC diffusion tubes used.
[c] Results presented as average of three samples, followed by the relative standard deviation (RSD) in parentheses.
Reference 129.

Ion chromatography has also been used in the monitoring of azide residues from discharged air bags.[131,132] Sodium azide is used explosively to inflate air bags with N_2 gas. The traces of azide ion that remain after the bag is used may constitute a health hazard for workers dismantling wrecked automobiles or replacing the azide charges. Some metal azides are also explosive. Figure 6.11 shows the separation of N_3^- ion from other ions.[133] Azide ion is usually only partly separated from bromide ion.

Several other methods of industrial hygiene/toxicology applications have been developed for the ion chromatograph. Acid gases such as HF and HCl can be run after these species are trapped in basic scrubbers.[134] Sulfate and oxalate ions have been found in a variety of industrial environments,[135] and ammonia and amines have been determined in environmental samples,[136,137] as has iodine.[138] Flame-retardant additives for polymers and the combustion products of polymers have been analyzed by IC.[139] This has been extended to various combustible components of aircraft interiors.[140] Figure 6.12 is an example of the toxic gases that are generated in these combustions.[141]

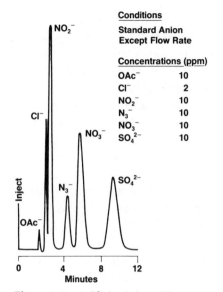

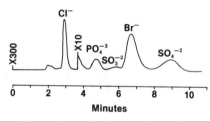

Figure 6.11. Azide in air-bag effluents. Reprinted with permission of *Ann Arbor Science.*

Figure 6.12. Toxic gas anions from combusted aircraft interior. Reprinted with permission of *Ann Arbor Science.*[140]

Forensic science uses of ion chromatography, only recently developed, include the determination of extractable ions in marijuana[142] and gasoline,[143] the analysis of explosives and explosive residues,[144] and soil extract analysis. Innovative work in this area should expand the use of ion chromatographs in the crime lab.

6.3 WATER POLLUTION

Water quality assessment and water pollution analysis are natural applications of ion chromatography.[145–158] Overall use of IC in these areas has been somewhat limited by the lack of method equivalence to the published methods of the Environmental Protection Agency and the American Public Health Association/American Water Works Association. It is necessary when working under the guidelines and regulations of these agencies to use equivalent methods. Once approval of IC is achieved, widespread acceptance of IC for these types of analysis is expected. IC is not currently useful in the analysis of many priority pollutants. Finally, in some cases IC offers only modest advantages over automated wet chemical methods.

Several characteristics of IC make it attractive for water analysis. One is the ion profile generated for each sample. The presence or absence of a wide variety of ions can be ascertained in a single injection on the ion chromatograph. Ions whose presence had not previously been known are detected as easily as those the analyst is searching for. In addition to the wide range of ions determinable by IC, another feature is the wide range of analyzable ion concentrations. The linear range for IC separations is generally 0.05 to 100 or 200 mg/L. Concentrations up to 2,500 ppm can be estimated with reasonable confidence, although the samples should be diluted for best results. The third aspect of IC analysis is the high sensitivity of the method for trace levels of ions, even when similar ions are present in much higher concentrations. One final feature is the wide separation of many chemically similar species such as $NO_2^- - NO_3^-$, the four halides, and various oxides of sulfur.

Drinking water from some cities in the United States has been analyzed for anion content (Figures 6.13 and 6.14). Standard anion IC instrument conditions are adequate to separate most of the common anions of interest in these samples. Table 6.5 is a summary of the results obtained by IC for a set of samples from the U.S. Geological Survey.[159] The USGS values are the result of an extensive long-term ion characterization by many analytical methods. Good agreement between IC and the "known" values

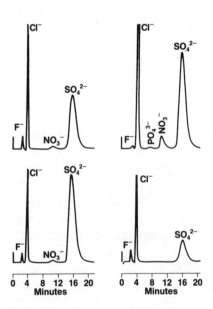

Figure 6.13. Ion chromatograms of drinking waters. Copyright by Dionex Corp. and reprinted by permission of copyright owner.

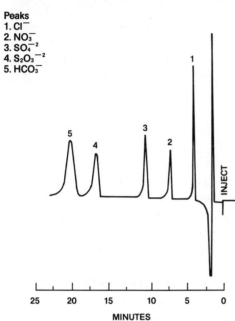

Conditions
Column: WESCAN Anion
Eluant: 0.004 m KHP, pH 4.5
Flow Rate: 2 ml/min
Sample: Tap water (Morgan Hill CA) — 100 μl
Detector: WESCAN Model 213 Conductivity Detector

Peaks
1. Cl^-
2. NO_3^-
3. SO_4^{-2}
4. $S_2O_3^{-2}$
5. HCO_3^-

MINUTES

Figure 6.14. Anions in drinking water. Courtesy of Wescan Corp.

Table 6.5. Comparison of IC Values with Reported Values of U.S. Geological Survey Reference Samples[a]

Sample number	USGS	IC	USGS	IC
	F^- (ppm)		Cl^- (ppm)	
54[A]	1.03	0.9	186	200
58[B]	0.92	0.5	1.7	1.4
60[C]	0.9	0.8	57	55
	NO_3^- (ppm)		SO_4^{2-} (ppm)	
54[A]	7.2	6.5	537	540
58[B]	NR	0.4	43.5	43
60[C]	NR	20.7	140	165

[A]Br^-—IC 0.4 ppm
[B]NO_2^-—IC 0.07 ppm
[C]Br^-—USGS 0.3 ppm/IC 0.3 ppm
   PO_4^{3-}—USGS 4.6/IC 3.8 ppm
   NO_2^-—IC 0.02 ppm

[a] Summary prepared by W. E. Rich, Dionex Corporation.   NR = Not Reported.

125

can be seen for the USGS samples. Anions in natural water sources have been analyzed and a thorough study in various types of water has been published.[160] In this work the overall precision of the IC method, estimates of method accuracy based on recovery studies from spiked samples, and comparison data from the ion chromatograh are presented. The ion chromatographic data compare favorably with the more established methods of ion analysis, although it is noted that sample throughput for IC for one ion is low compared with the one-ion-at-a-time methods. No comparison of overall throughput is made.

One important use of IC is the analysis of nutrients necessary for microorganism growth in surface waters.[161] In most samples the bulk of the micronutrients is present in ionic form. Ammonium, nitrate, orthophosphate, and sulfate are easily analyzed by IC. Nitrite and the polyphosphates are not generally determinable by IC at the naturally occurring levels of these ions. Total nitrogen-N and phosphorous-P analyses still require chemical digestion before analysis.

Several other water-quality parameters can be tested by IC. Water hardness constituents (Ca^{2+} and Mg^{2+}) can be determined by divalent cation analysis.[162] Total ionic content (TIC) can also be rapidly determined by IC. This parameter is directly related to total dissolved solids (TDS) in samples. The TIC test is much faster than the TDS. An ion chromatograph has been used in at least one industrial plant to monitor TIC.[163] Water-formed deposits and corroded materials have also been analyzed by IC.[164,165] Sodium methanearsonate has been determined in raw and biologically treated waste waters.[166]

An area in which IC is more frequently used is in the characterization of pore waters,[167,168] geothermal waters,[169] and ocean brines.[170–173] Sediment cores taken from rivers, lakes, and ocean bottoms naturally contain water trapped by the solid materials. The level of ionic content in these waters is quite high. IC has been used to analyze a variety of anions in the entrapped pore waters. Brine samples have also been analyzed by ion chromatography. The IC is suitable for analysis of high chloride samples as long as the Cl^- ion concentration is less than about 5000 mg/L (0.5%). For samples with higher chloride levels sample dilution or treatment with a silver precolumn to remove most of the chloride is necessary. Each of these methods makes trace bromide analysis difficult because dilution results in a bromide concentration too low to be detected by IC and silver pretreatment removes the bromide as well as the chloride. Other ions such as NO_3^-, PO_4^{3-}, and SO_4^{2-} can be analyzed after the silver pretreat-

Standard Anion

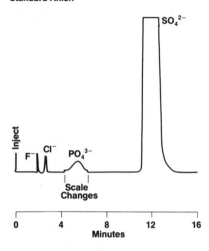

Figure 6.15. Anions in ocean water after chloride removal. Copyright by Dionex Corp. and reprinted by permission of copyright owner.

ment (Figure 6.15).[174] The newly introduced SA-2 anion-separating column has been successful in separating trace Br^- in high Cl^- because its elution profile is quite different from that of the normal (SA-1) anion separator. Bromide and nitrate elute *after* sulfate with this new resin.[175]

Manufacturing industries need large quantities of water. Industrial uses include heating and cooling, dissolving or transporting raw materials, diluting intermediates or finished products, and cleaning finished products. Industrial uses of IC range from analyses of micronutrients in influent waters, analysis of process stream waters, and cooling tower waters to the final plant effluents that are dumped back into rivers and lakes as outfall.

The ion content of effluents is regulated by various government agencies. The levels of the regulated ionic species must be routinely checked to determine whether the plant is in compliance with its National Pollution Discharge Elimination System (NPDES) permit. Waste streams are often chemically or biologically treated to reduce the levels of contaminants to meet discharge requirements. Biopond waters are analyzed after bacteria have acted on the waste. Deviations in the content of the waste entering the biopond can damage or destroy the bacteria colony. In one case history acetate and formate ions were monitored because bacteria in one particular treatment pond were susceptible to these acids.[176] For these samples UV radiation was used to kill the bacteria before IC injection.

This was necessary because the bacteria consumed the resins. Chemical treatment such as chlorination and flocculation is also effective in reducing concentrations of some ions. Equipment failures at chemical treatment facilities can sometimes be located by IC analysis. Two examples are (1) the determination by the presence of high SO_4^{2-} values when an H_2SO_4 treatment process involving ion exchangers is not working properly because the resin beds are saturated and (2) glycolate ion deficiencies in boiler blow-down waters indicate that insufficient EDTA is being pumped into the boiler (i.e., the pump has failed).[177]

One of the most important uses of IC in water pollution analysis is the determination of ion content in final effluent waters (Figures 6.16 and 6.17).[178,179] The feasibility of applications has been demonstrated in monitoring ions in green pulping liquors and Kraft black liquors. Usually only dilution and filtration are required as sample pretreatments. Table 6.6 summarizes the other applications of IC in the pulp and paper industry. Figures 6.18 and 6.19 illustrate an ion chromatogram of a typical Kraft black liquor sample[180,181] in which the oxalate and sulfate ions are well separated and easily quantitated. Another use of IC is in coal liquefaction and gasification processes[182] and oil shale/tar sand extracts.[183]

The use of IC as a routine analysis tool in water quality work is presently limited by the lack of equivalency of the method. Once the IC is designated as an equivalent method there should be a jump in growth of the use of IC in water pollution analysis.

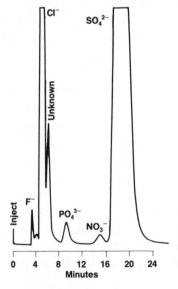

Figure 6.16. Anions in wastewater. Copyright by Dionex Corp. and reprinted by permission of copyright owner.

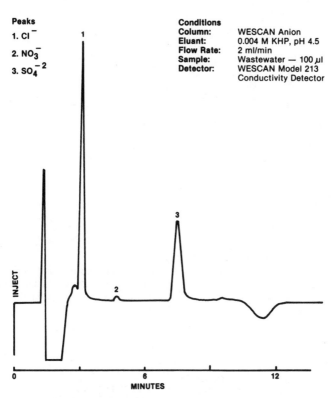

Peaks

1. Cl$^-$
2. NO$_3^-$
3. SO$_4^{-2}$

Conditions

Column:	WESCAN Anion
Eluant:	0.004 M KHP, pH 4.5
Flow Rate:	2 ml/min
Sample:	Wastewater — 100 μl
Detector:	WESCAN Model 213
	Conductivity Detector

INJECT

0   6   12

MINUTES

Figure 6.17.   Anions in wastewater by single-column IC. Courtesy of Wescan Corp.

Table 6.6.   Uses of IC in the Pulp and Paper Industry

Sample Type	For	Concentrations
Hatchery water	NO_3^--N	20–60 ppb
	PO_4^{3-}-P	5–20 ppb
Paper mill effluent	SO_4^{2-}-S	10–100 ppb
Biopond water	NO_3^--N	10–50 ppb
	PO_4^{3-}-P	100–175 ppb
KCl soil extract	SO_4^{2-}-S	100–300 ppm
Acetate soil extract	SO_4^{2-}-S	10–100 ppm
Ammonium fluoride soil extract	PO_4^{3-}	1–10 ppm
Sulfuric acid soil extract	PO_4^{3-}	10–50 ppm
Bicarbonate soil extract	SO_4^{2-}	1–5 ppm
Pulping liquid (Green)	S^{2-}	Very high
	SO_4^{2-}	Very high

Reference 174.

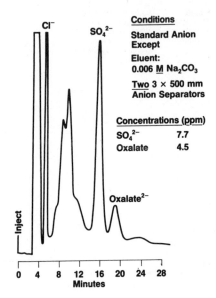

Conditions

Standard Anion Except

Eluent:
0.006 M Na₂CO₃

Two 3 × 500 mm **Anion Separators**

Concentrations (ppm)

SO₄²⁻	7.7
Oxalate	4.5

Figure 6.18. Sulfate and oxalate in Kraft black liquors. Copyright by Dionex Corp. and reprinted by permission of copyright owner.

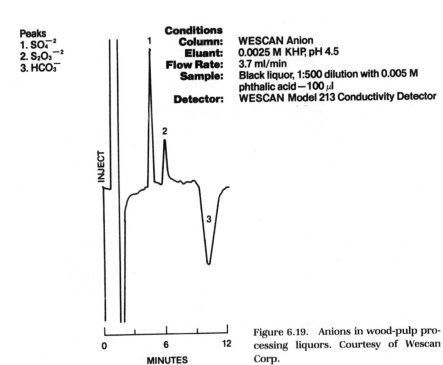

Peaks
1. SO₄⁻²
2. S₂O₃⁻²
3. HCO₃⁻

Conditions

Column:	WESCAN Anion
Eluant:	0.0025 M KHP, pH 4.5
Flow Rate:	3.7 ml/min
Sample:	Black liquor, 1:500 dilution with 0.005 M phthalic acid — 100 µl
Detector:	WESCAN Model 213 Conductivity Detector

Figure 6.19. Anions in wood-pulp processing liquors. Courtesy of Wescan Corp.

6.4 INDUSTRIAL APPLICATIONS

A diverse application area in which IC has found use is in the solution of problems in industrial ion analysis.[184-186] In addition to the final effluent, cooling water, and commercial product analyses covered elsewhere in this chapter, ion chromatography has been applied to industrial research, competitive product analysis, process monitoring, and quality control. Other uses include generating ion profiles and materials balances for final products and intermediates and the analysis of impurities of precursors or feedstock materials. Many IC methods involve proprietary processes or products. These necessarily are not discussed in this chapter.

To facilitate the documentation of the industrial uses of IC the following arbitrary division of industries has been devised:

1. Petroleum.
2. Polymer processing and products.
3. Chemical manufacturing.
4. Automobile and steel.
5. Mineral and paper.
6. Electronics industry.

The power production/energy uses of IC have become large enough to be classed as a separate section. Commercial formulations and products, food, beverage, and food additive applications are also covered in a separate section.

The petroleum processing industry has developed some major applications for IC analysis, several of which are related to air pollution and water pollution analysis. Other analyses are specifically related to the petroleum industry.

One major application is the IC determination of oil-field brine tracers.[187] Brines are injected into oil fields to enhance oil recoveries. These NaCl solutions have a wide range of salt concentrations (2 to 25%). The brines are heated, heavily doped with ionic and nonionic surfactants and other additives, and pumped into the oil field. Ionic tracers are used to follow the flow of the brine through the field. Four common anions are used as tracers: bromide, nitrate, iodide, and thiocyanate. Samples are usually only diluted just before injection into the IC for the determination of these ions. Figure 6.20 illustrates the detection of I^- and SCN^- ions in brines. The presence of surfactants and other additives does not interfere with the determination of these anions.

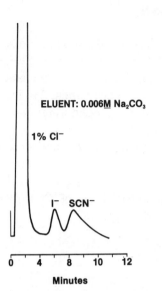

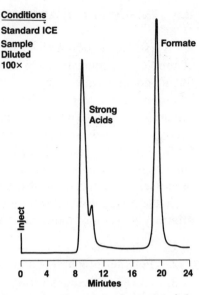

Figure 6.20. Iodide–thiocyanate separation in 1% NaCl. Copyright by Dionex Corp. and reprinted by permission of copyright owner.

Figure 6.21. Formate and acetate in brine by ICE. Copyright by Dionex Corp. and reprinted by permission of copyright owner.

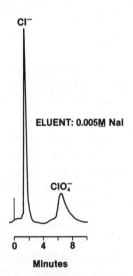

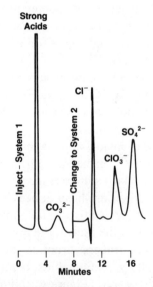

Figure 6.22. Perchlorate in brine. Copyright by Dionex Corp. and reprinted by permission of copyright owner.

Figure 6.23. Separation of anions in 4% caustic. Reprinted with permission of *Ann Arbor Science*.[194]

Several other ions are analyzed in brines. Sulfate, calcium, and magnesium are determined because these ions can result in corrosion deposits. Formate has been determined in a 25% $MgCl_2$ brine solution by anion ICE. Perchlorate ion has also been determined in brine with $0.005M$ NaI eluent and silver suppression.[188] Figures 6.21 and 6.22 are typical examples.

One important limitation in IC analysis of brines is that the NaCl concentration in the injected sample must be less than 1%; if higher NaCl concentrations are injected, column overloading will result. Previously, L-20, brine, or GLL-2 columns (all names for the same resin) were used to enhance the I^-–SCN^- separation. Now, use of p-cyanophenol with standard columns or use of SA-2 separator columns (possibly a similar resin) may have eliminated the need for these special resins.

IC has also been used for several other petroleum applications. Sulfur has been determined in combusted samples of gasoline and diesel fuel,[189–191] anions in refinery process waters,[192] and oil well cements[193] have been analyzed for anion and cation content.

Polymer processing scientists have several analytical requirements that are met by ion chromatography. A major IC application in polymer processing is the analysis of caustic (NaOH) solutions.[194] The ions of most frequent interest in these solutions are CO_3^{2-}, Cl^-, ClO_3^-, and SO_4^{2-}. Carbonate ion is difficult to analyze under standard anion IC instrument conditions. Use of coupled IE/IC minimizes the influence of OH^- ions from the sample acting as eluting ions and minimizes interference between chloride and carbonate ions.[195] Figure 6.23 shows an IE/IC separation of the ions of normal interest from a standard of 4% NaOH. Problems have been encountered when the NaOH concentration approached 30 to 50%. Although the carbonate peak remains constant, the void volume peak that contains the chloride, chlorate, and sulfate ions shows substantial increases after repeated injections. Subsequent anion IC analysis of Cl^-, ClO_3^-, and SO_4^{2-} also shows these increases; yet injection of standards that contain only Cl^-, ClO_3^-, and SO_4^{2-} immediately after high concentration caustic injections show peak heights exactly like earlier runs. No explanation has been suggested for this observation.[196]

Several other IC analyses have been developed for direct treatment of polymer products and impurities. One good example involves methacrylic acid.[197] In this case the sample is dissolved in a mixture of acetone/H_2O and injected into the ion chromatograph. The methacrylate ion can be assayed and trace levels of hydroxybutyric acid impurities can be measured at the same time. Another use for IC is the analysis of sulfate in certain polymer feedstocks.[198,199] Simple dilutions and standard IC analyses are usually adequate to produce sulfate data. Sulfate ion is a possible cause of corrosion and can poison some catalysts. Phenolic polymers have been dissolved in toluene and then extracted with water or eluent.[200] Subsequent analysis by conventional anion IC showed a variety of anions in the sample. Fluoride in nylon polymers has also been determined by IC.[201]

The chemical manufacturing and processing industries use IC in a number of ways. Guanadine, nitroguanadine,[202] and a variety of phosphonic acids[203] (Figure 6.24) have been analyzed in explosives mixtures. Organophosphoric and organophosphothioic acids have been analyzed in aqueous streams.[204] Fertilizers have been analyzed with some success for NH_4^+, Ca^{2+}, Mg^{2+}, NO_3^-, PO_4^{3-}, and SO_4^{2-} ions,[205] and tripolyphosphate and pyrophosphate have been separated with $0.01M$ sodium citrate as the eluent (Figure 6.25).[206] However, detector and column fouling have hampered extensive use of the reported separations method.[207] Impurities in surfactants have been analyzed by IC with $0.0015M$ NaHCO₃ to elute

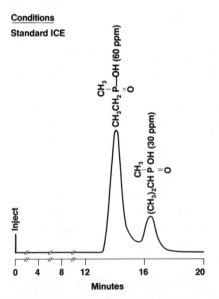

Figure 6.24. Separation of phosphonic acids. Reprinted with permission of *Ann Arbor Science*.

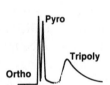

Figure 6.25. Separation of ortho-, pyro-, and tripolyphosphates.

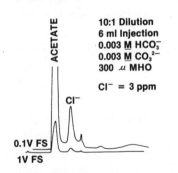

Figure 6.26. Chloride impurities in acetic acid. Reprinted with permission of *Ann Arbor Science*.[210]

glycolate, chloroacetate, and chloride ions.[208] Nitrite and nitrate ions are easily analyzed in high concentration H_2SO_4.[209] Chloride impurities in acetic acid (Figure 6.26)[210] and strong anions in ethyl acetate have also been determined by ion chromatography.[211] Anions in inks, dyes, cutting fluids, and other commercial products have been determined by IC and IE/IC methods.[212] Anions in photographic developing solutions have also been determined by IC (Figure 6.27).[213–215] Finally, amine impurities in dimethylformamide have been determined under standard cation IC conditions (Figure 6.28).

The automobile and steel industries utilize IC analysis in many ways. Exhaust emissions are easily monitored by IC, and ethylene glycol antifreezes are routinely monitored for decomposition of additives. Nitrite, benzoate, phosphate, sulfate, and glycolate are all determined in these

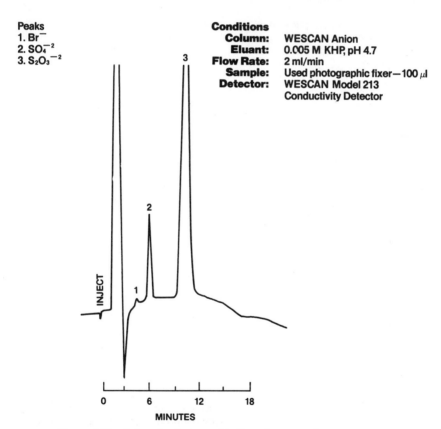

Peaks
1. Br^-
2. SO_4^{-2}
3. $S_2O_3^{-2}$

Conditions
Column: WESCAN Anion
Eluant: 0.005 M KHP, pH 4.7
Flow Rate: 2 ml/min
Sample: Used photographic fixer—100 μl
Detector: WESCAN Model 213 Conductivity Detector

Figure 6.27. Anions in photographic fixer. Courtesy of Wescan Corp.

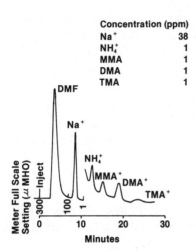

Concentration (ppm)	
Na$^+$	38
NH$_4^+$	1
MMA	1
DMA	1
TMA	1

Figure 6.28. Amine impurities in DMF. Copyright by Dionex Corp. and reprinted by permission of copyright owner.

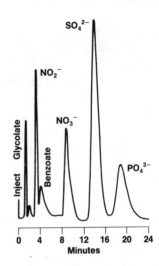

Figure 6.29. Anions in ethylene glycol. Copyright by Dionex Corp. and reprinted by permission of copyright owner.

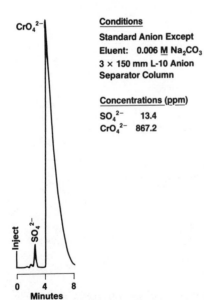

Conditions

Standard Anion Except

Eluent: 0.006 $\underline{M}$ Na$_2$CO$_3$

3 × 150 mm L-10 Anion Separator Column

Concentrations (ppm)

SO$_4^{2-}$	13.4
CrO$_4^{2-}$	867.2

Figure 6.30. Chromate and sulfate in copper plating bath. Copyright by Dionex Corp. and reprinted by permission of copyright owner.

samples.[216] Figure 6.29 is a typical separation of these ions. Plating baths have been analyzed for fluoride, chloride, and sulfate content.[217,218] Chromate, gold complexes and other ions have also been determined in certain plating baths (Figures 6.30 and 6.31). Problems have been encountered when certain organic plating bath additives have interferred with rapidly eluting ions. Use of p-cyanophenol in the eluent should allow routine determination of CrO_4^{2-} without the use of L-10 or L-20 separating columns. In a similar application pickling baths required in specialty steel processing are easily monitored for stong acid content by IC.[219]

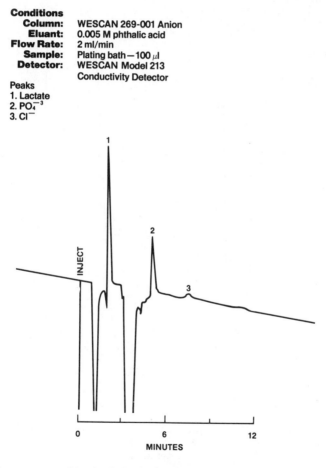

Conditions
Column: WESCAN 269-001 Anion
Eluant: 0.005 M phthalic acid
Flow Rate: 2 ml/min
Sample: Plating bath — 100 μl
Detector: WESCAN Model 213
Conductivity Detector

Peaks
1. Lactate
2. PO_4^{-3}
3. Cl^-

Figure 6.31. Anions in plating bath solution. Courtesy of Wescan Corp.

A few IC applications related to mineral and paper processing have been developed. One major use is in the aluminum industry's analyses of Bayer liquors.[220,221] Sulfate and oxalate are routinely determinable in these samples. Figures 6.32 and 6.33 show that both ions are in fairly high concentration and well separated. Once again a problem has been noted in the suppressor-type anion separator columns used in this application. Periodically the separating column must be soaked overnight in a NaOH

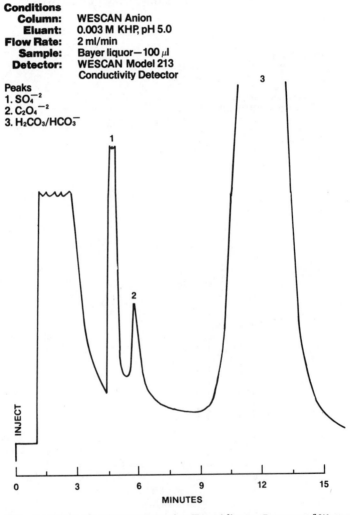

Conditions
Column:	WESCAN Anion
Eluant:	0.003 M KHP, pH 5.0
Flow Rate:	2 ml/min
Sample:	Bayer liquor—100 μl
Detector:	WESCAN Model 213 Conductivity Detector

Peaks
1. SO_4^{-2}
2. $C_2O_4^{-2}$
3. H_2CO_3/HCO_3^-

INJECT

MINUTES

Figure 6.32. Anions in aluminum-processing (Bayer) liquor. Courtesy of Wescan Corp.

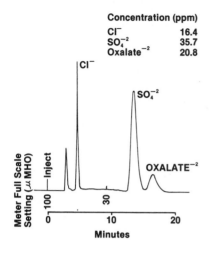

Figure 6.33. Suppressor-type anion determination of Bayer liquor. Copyright by Dionex Corp. and reprinted by permission of copyright owner.

solution (0.2 to 0.5M) to restore resolution between oxalate and sulfate ions.[222] This treatment is *not* recommended for silica columns because it is not known whether the same problem occurs and because the NaOH would destroy the column packing. These ions are also determined in Kraft black liquors in the pulp and paper industry. Other analyses done on similar but less complex samples are biopond waters, soil extracts, and mill effluents.[223]

Nuclear fuel reprocessing has developed one unusual IC application. Uranium is removed from fission products and cladding materials by treating them with tributylphosphoric acid/kerosene mixtures. The tributylphosphate forms a uranyl complex that breaks down in weak HNO_3. Monobutyl- and dibutylphosphate impurities lower the yield of recovered uranium by forming complexes that do not break down in HNO_3. Monobutyl- and dibutylphosphates have been determined in an NaOH cleanup solution used to treat the extracting solution.[224–226]

The electronics industry has begun to apply IC in a variety of ways. Ion chromatographs are used to analyze plating baths, measure impurities in deionized water, and determine anions as surface contaminants on aluminum and zinc metals[227] in components manufacturing. Circuit-board impurities and encapsulation plastics (Figure 6.34) have been studied with water[228] and isopropyl alcohol washes.[229] Both quaternary ammonium compounds and inorganic salt battery additives have been measured by IC,[230] and general problem solving with IC an an ion analysis method has been reported.[231,232]

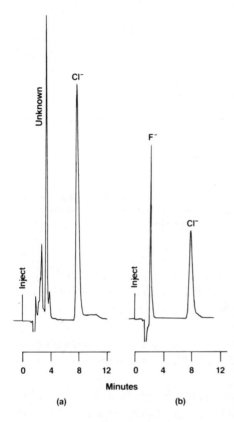

Figure 6.34. (a) Anion profile in encapsulate plastic extract; (b) Anion standard. Copyright by Dionex Corp. and reprinted by permission of copyright owner.

A variety of industrial uses for ion chromatography is under development. Sulfur in industrial steel[233] has been analyzed by IC, which has also been used in geochemical exploration.[234–239] Unfortunately many industrial uses of IC are proprietary and cannot be included in this text. Because IC can handle the complex nature of the matrices in many industrial samples in a straightforward way, uses of IC by industry should proliferate. Improved sample pretreatment methods and column preservation and rejuvenation methods should aid in this development.

6.5 ENERGY/POWER PRODUCTION

One industry that has embraced ion chromatography to solve some of its difficult analytical problems is the electricity generating industry. One major application that has been developed for IC is the determination of

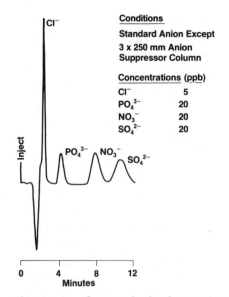

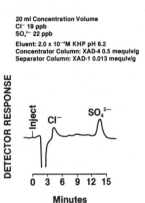

Figure 6.35. Ultratrace levels of anions in power plants. Copyright by Dionex Corp. and reprinted by permission of copyright owner.

Figure 6.36. Trace detection of chloride and sulfate by single-column IC. Reprinted from *Analytical Chemistry* with permission.[244] Copyright 1981 American Chemical Society.

ultratrace (μg/L or ppb) levels of chloride, sulfate, and sodium ions in steam condensate, polishing bed effluents, and other ultraclean waters.[240–250] Detection limits below 1 ppb have been calculated for chloride and sulfate. These ions contribute to metal stress and corrosion in water-contacting electrical equipment. Because power plants are operated at high temperatures and pressures, the corrosion and metal stress caused even by μg/L levels of these ions can be severe.

Figures 6.35 and 6.36 show chromatograms of ppb levels of anions run on standard anion system.[251] Under normal operating conditions the ion chromatograph has detection limits between 10 and 100 ng with 0.1 mL loop injections. Concentrating columns were developed to enhance detectability for ultratrace work. These columns concentrate the ions from the sample to allow detection of 0.1 to 1 ng/mL in the sample. Note that the actual detection limits for IC have not been lowered but that the samples have been concentrated up into the detector range of IC. Typically, 10 to 50 mL of clean sample are used to bring the ion concentrations into the detection range of the ion chromatograph. These columns have been adopted for general use in the analysis of steam condensates and other ultrapure waters. Recent work indicates that if loading volume is held constant detection of 1 to 10 μg/L of Cl^- and SO_4^{2-} in the presence of 1 to 2 ppm NH_3 is achievable. If the loading volume varies, however, the chloride response will vary nonlinearly.[252] Use of a higher capacity 3 × 150 or 4 × 50 mm column will help to solve this problem.[253] Large volume loop injection directly onto the separating column will also reveal some cations at these low levels. This method works well to detect 10 ppb Na in the presence of 2 to 3 ppm NH_3.[254]

Other power-plant uses of IC include the analysis of influent and cooling tower waters, the determination of ions in corrosion deposits and sulfur in coal and heavy fuel oil, and stack gas and flue gas desulfurization feedstream analysis. Morpholine has been reported to elute after potassium under cation IC conditions, although column conditioning with high morpholine concentration may be required. Borate ion has been determined in several types of sample with ICE-based elution.[255] Finally, hydrazine has been detected by electrochemical detectors (Figure 6.37).[256]

Several IC applications have been developed to meet the special demands of nuclear power industry. Trace levels of chloride in high boric acid/borate solutions and in the presence of 10 to 50 ppm of iodide have

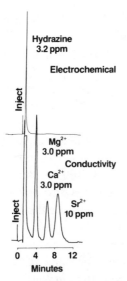

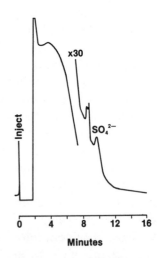

Figure 6.37. Determination of hydrazine. Copyright by Dionex Corp. and reprinted by permission of copyright owner.

Figure 6.38. Anions in fused and dissolved glass.[258]

been detected.[257] Numerous nuclear plants are now being equipped with a modified ion chromatograph to perform routine power-plant analyses and ultratrace chloride analyses under TMI-type emergency conditions.

IC has also been used in the study of nuclear waste disposal. One proposed method for waste containment encapsulates the radioactive waste in glass.[258–262] Sulfate, nitrate, and phosphate ions have been studied on a specially modified ion chromatograph. The glass samples are first fused with tetraborate or carbonate, and the resulting pellets are dissolved in HCl and injected into the IC for analysis. Figure 6.38 is a typical ion chromatogram for this analysis.

Several other IC uses related to energy production include fuel cell effluents,[263] quaternary ammonium battery additives, and zinc and lithium salts used in high efficiency batteries. So far ion chromatography has not been successful in separating or detecting large phosphonate-type boiler water additives.

6.6 FOODS, BEVERAGES, AND FOOD ADDITIVES

Use of ion chromatography is just starting in the determination of ions in foods, beverages, and food additives. Although not yet in general use in the food industry because the cost per analysis is fairly high compared with existing analyses, numerous interesting IC food applications have been demonstrated.

Several types of foodstuff have been analyzed for ionic content. Barley has been run to measure levels of NO_2^- and NO_3^- ions.[264] First a weighed portion of grain is extracted in anion eluent in a blender. After filtration standard anion IC conditions with small suppressors showed no nitrite and only a trace amount of nitrate. Soy protein has also been analyzed for nitrite ion.[265] Here ICE standard conditions were used. Meat samples have been run on IC, with sulfate, nitrite, and nitrate as the ions of interest. In uncured meats these same ions are analyzable. In cured meats, however, large amounts of organic ions make these determinations difficult.[266]

IC has been much more extensively used in beverage analysis. Most of these products contain numerous common inorganic ions as well as a variety of carboxylic acids. Figures 6.39 to 6.43 are examples of ions in red wine, milk, coffee, and colas.[267–271] Sugar solutions and fermentation broths have also been analyzed by IC.

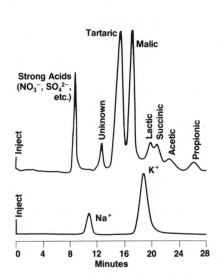

Figure 6.39. Red wine acids and cations. Copyright by Dionex Corp. and reprinted by permission of copyright owner.

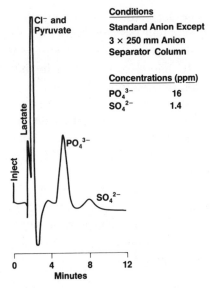

Figure 6.40. Common anions in milk. Copyright by Dionex Corp. and reprinted by permission of copyright owner.

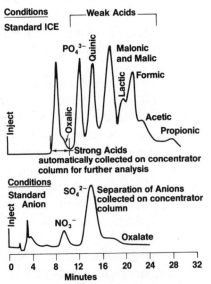

Figure 6.41. Strong and weak acids in coffee extracts. Copyright by Dionex Corp. and reprinted by permission of copyright owner.

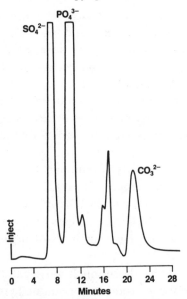

Figure 6.42. ICE analysis of cola beverage. Copyright by Dionex Corp. and reprinted by permission of copyright owner.

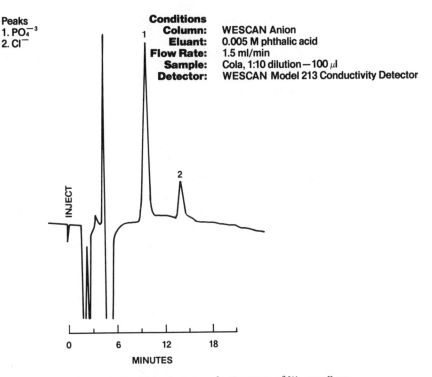

Peaks
1. PO_4^{-3}
2. Cl^-

Conditions
Column:	WESCAN Anion
Eluant:	0.005 M phthalic acid
Flow Rate:	1.5 ml/min
Sample:	Cola, 1:10 dilution—100 μl
Detector:	WESCAN Model 213 Conductivity Detector

Figure 6.43.   Phosphate in cola. Courtesy of Wescan Corp.

Several food additives have been analyzed for ion content, and both cyclamate and saccharine artifical sweeteners have been examined for anion content.[272] Sulfite is added to shrimp to prevent oxidation of the meat. Levels of SO_3^{2-} are regulated by several nations. Sulfite, sulfate, and phosphate ions have been determined in extracted shrimp.[273] These ions have also been determined in pharmaceuticals.[274] Table salt has been investigated by IC for SO_4^{2-} and organic acid content,[275] and low levels of both types of ion have been determined in fairly high concentrations of brine. Prussate of soda, used in salt purification, has been analyzed by IC. Finally, food dyes are regulated by the Food and Drug Administration for anion content.[276,277] Ion chromatography has been used successfully on the determination of chloride, sulfate, and iodide in these samples. Large macromolecule coloring agents have also been analyzed for acetate and formate impurities by IE analysis.[278]

6.7 PERSONAL AND COMMERCIAL PRODUCTS

Numerous commercial products and personal-use items have been analyzed by IC. Many of these analyses have been done to uncover impurities
that affect product performance.

Paints and surfaces about to be painted have been analyzed by IC. In
one paint sample excessive lactic acid was found in a formulation that
did not cure well. In another case automobile bodies stored before being
painted showed blisters when the paint was applied.[279] The liquid from
the blisters was removed for IC analysis and it was found that acid mists
had deposited H_2SO_4 during storage. Swimming pool chemicals have been
analyzed for oxides of chlorine.[280] In this type of sample the ClO^- and
Cl^- are not separable by normal anion IC methods. In one suggested
method $S_2O_3^{2-}$ was reacted with the ClO^- ion and the residual peak was
measured as chloride. Use of the electrochemical detector to determine
ClO^- is a straightforward procedure (Figure 6.44).[281] Detergents and fertilizers are easily analyzed for a variety of ions which includes PO_4^{3-},
SO_4^{2-}, NO_3^-, Cl^-, and NH_4^+.[282] Pyrophosphate and tripolyphosphate
have recently been separated,[283] although difficulties with the detector
and resins have been reported. Nitrilotriacetic acid and ethylenediamine
tetracetic acid have not yet been separated and detected by IC. Cutting
fluids have been analyzed for oxalate and sulfate ions.[284] Ionic impurities

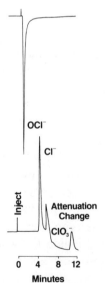

0   4   8   12

Minutes

Figure 6.44.   Fresh commercial bleach. Copyright by Dionex Corp.
and reprinted by permission of copyright owner.

in ink formulations, surfactants (Figure 6.45),[285] and pesticides[286] have been determined by IC. The high concentrations of organic materials in these samples either pass through in the void volume of the column set or are absorbed in the precolumn. Periodic treatment of the precolumn to remove these species must be done. Cement samples have also been run for anions and cations.[287]

Numerous personal products have been analyzed for ions by IC. Tobacco extracts have been run for fluoride and phosphate ions.[288] Ions in dyes used in cosmetics[289] and alkanolamine content in hair dyes and other products can be analyzed by ion chromatography,[290] and fluoride, monofluorophosphate, and orthophosphate have been determined in toothpaste samples.[291] Figure 6.46 is a typical chromatogram for these three species. Another anion separator column problem has been noted in conjunction with this analysis. The values obtained for free fluoride ions develop a high bias after samples have been run for several months. Overnight treatment of the column in $12M$ H_2SO_4 at two-to-three-week intervals seems to delay this problem. The presence of silicates and other fouling species are posed as tentative explanations for this difficulty. Mouthwashes that contain fluoride ion are also easily analyzed by IC.[292] Finally, vitamins and carbenacillin-type pharmaceuticals are assayed by IC.[293]

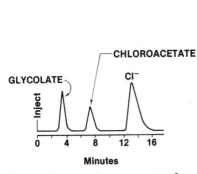

Figure 6.45. Anionic components of a surfactant mixture. Copyright by Dionex Corp. and reprinted by permission of copyright owner.

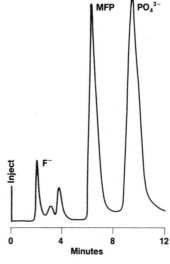

Figure 6.46. F^-, MFP, and PO_4^{3-} in toothpaste. Courtesy of Colgate-Palmolive.[291]

The use of IC in personal-product and commercial formulation analysis has become quite broad. It is hoped that other interesting uses of IC in these types of sample will be reported in the near future.

6.8 MICROELEMENTAL ANALYSIS

The ion chromatograph was applied early to the determination of heteroatoms in solid samples. Heteroatoms successfully analyzed by IC include sulfur, chlorine, bromine, phosphorous, and iodine.[294–296] Solid samples are broken down by combustion or digestion before IC analysis. Combustion is the most widely used method of degrading samples and is most often done in a Schoeniger flask. Small quantities of sample (1 to 5 mg) are carefully weighed and burned in this low-pressure oxygen environment. If trace levels are being analyzed, if the sample burns poorly, or if residues are left after combustion, an oxygen bomb should be used. Acid digestion, oxidation by persulfate or perchlorate, or base hydrolysis are done if combustion procedures are inadequate to convert all the heteroatoms to one ionic form.

Combustion or digestion converts the heteroatoms to gases directly or to anionic or cationic forms. These are collected by a scrubbing solution that forces all the possible species of one heteroatom into a single form. After trapping the heteroatoms in the scrubbing solution, the solution is injected directly into the ion chromatograph. Both oxidizing and reducing scrubbers have been used in conjunction with IC analysis. The most commonly used scrubber is a mixture of hydrogen peroxide with standard anion eluent. A typical chromatogram for a combusted sample that uses this scrubber is shown in Figure 6.47. The large peak results from the generation of CO_2 gas[297] during combustion of the carbon in the sample or from sodium breaking through the suppressor bed. The IE/IC-coupled chromatographic system can be used to eliminate any potential Cl^-/ CO_3^{2-} interference or to eliminate Na^+ breakthrough. Another scrubber is a hydrazine solution,[298] used to shift the nitrogen combustion products from NO_3^- to the reduced NO_2^- form. This is important in trace bromide analysis in which NO_3^- in large amounts would swamp the small Br^- peak.

Samples combusted by the simple Schoeniger method have been analyzed for sulfur and chlorine content. For the analysis of chlorine there is a problem when the sample weight is below 1 mg. It may be due to

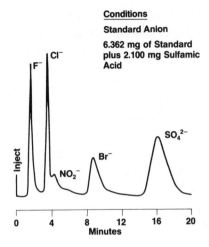

Conditions

Standard Anion

6.362 mg of Standard plus 2.100 mg Sulfamic Acid

Figure 6.47. Combusted multiple halogen standard mixed with sulfamic acid. Reprinted from the *Microchemical Journal* with permission.[295]

CO_3^{2-}/Cl^- interference, to the "water dip" because the sample was run with a large suppressor column, or to homogenity in the sample. Use of larger amounts of sample results in good agreement between IC and other methods. Trace quantities of sulfur in solid samples are best run by IC. Table 6.7 lists the results for low levels of sulfur in solid samples. For most samples less weight is required to do the analysis by IC than by conventional methods.[299]

One heteroatom that is not easily analyzed by IC is nitrogen.[300] Low recoveries from combustions of standards may indicate that nitrogen is converted to N_2 gas as well as to NO and NO_2. This problem occurs even when a highly pressurized oxygen bomb is used. Use of a different scrub-

Table 6.7. Comparison of Sulfur by Ion Chromatography (IC) to Barium Precipitation Techniques (BPT)

Sample	Weight for IC (mg)	IC	BPT
741	5.611	9.61	9.70
749	5.242	12.02	12.10
868	5.447	12.09	11.94
295	8.262	4.51	4.60
641	6.947	3.88	3.90
642	1.041	4.66	4.81

Reference 299.

ber has resulted in efficient NO_x trapping and much better recoveries.[301] Successful nitrogen analyses by IC, done after the samples are digested, convert all the nitrogen to NH_4^+. This ion is easily determined by standard monovalent IC analysis.

In addition to pure synthesized compounds, the ion chromatographic analysis of heteroatoms has been used routinely in a number of areas. Toxic gases generated from burned aircraft components have been studied.[302] Other polymers and flame-retardant additives have been analyzed by IC.[303] Coal, gasoline, and diesel fuels have been analyzed for sulfur content,[304] and iodide in volatile samples has been determined.[305]

6.9 BIOMEDICAL RESEARCH

One area in which IC has only begun to be used is biomedical research. Some interesting applications have been developed for suppressor-type IC. These methods, however, have not been widely applied to routine clinical analyses. Anion and monovalent cations have been separated in blood, serum, and urine.[306] Oxalate ion has been determined in urine samples by using longer separator columns to get better oxalate–sulfate separation.[307] Oxalate has also been directly determined in serum, and samples of bromide in blood and serum have been run on selected groups of industrial workers. A feasibility study of using IC and coupled ICE/IC to analyze pyruvate and lactate in serum has also been done.[308] Investigation of urinary vanillylmandelic acid in the same study concluded that electrochemical detection results in better sensitivity of this species. It is also stated that reverse-phase silica columns can probably take the place of the ICE columns in coupled chromatographic separation. Organic acids in human fluids[309] and calcium and magnesium ions in serum[310] have also been determined by IC. Impurities in drugs have been screened by IC and hypoalamentation and intravenous solutions have been analyzed for a number of ions.

Other uses related to physiological analyses have been developed. Plant extracts and animal feeds have been run on IC to determine the presence of certain ions,[311] and sulfate has been analyzed in biological fluids.[312] The use of IC in biological and biomedical research applications is beginning to expand.

6.10   OTHER RESEARCH USES OF IC

This section covers research that has been done in IC to expand into new areas. It often involves use of other detectors after conventional conductivity detection has shown inadequate results. Other column packings may be substituted or developed when the resins used in conventional IC do not give adequate separation.

Early basic research already incorporated into the methodology of ion chromatography includes coupled chromatography,[313] azide analysis,[314,315] and kinetic studies of SO_3^{2-}–SO_4^{2-} conversion. Other methods not yet widely used in IC analyses are the resistivity detection of weak ions (Figure 6.48),[316,317] analysis of transition metals as their tartrate complexes (Figure 6.49), [318] use of a coulometric detector for cyanide determination,[319,320] and use of a photometric detector to analyze antifreeze additives[321] and sulfur compounds.[322] Other research efforts include studies of the slow oxidation of halogenated hydrocarbons,[323] transition metal determinations with barium eluents and sulfate form suppression[324] and lead eluents with iodate suppression to show that the cation-separating resin had insufficient selectivity to separate many transition metals,[325]

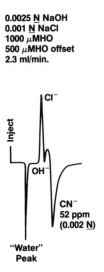

0.0025 N NaOH
0.001 N NaCl
1000 μMHO
500 μMHO offset
2.3 ml/min.

Cl⁻

Inject

OH⁻

CN⁻
52 ppm
(0.002 N)

"Water"
Peak

Figure 6.48.   Resistivity detection of weak anions. Reprinted with permission of *Ann Arbor Science*.[316]

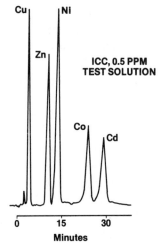

Cu

Ni

Zn

ICC, 0.5 PPM
TEST SOLUTION

Co

Cd

0   15   30

Minutes

Figure 6.49.   Transition metal complexes by Coulemetric detector. Reprinted from *Analytical Chemistry* with permission.[319] Copyright 1979 American Chemical Society.

and a study of numerous sulfur species with conductivity and photometric detectors.[326] Iron interferences have also been removed by complexation.[327] Recent work on mobile-phase ion chromatography (MPIC),[328] a method very similar to paired ion liquid chromatography, and high-speed[329] and high-performance[330] ion chromatography has been reported. MPIC has separated surfactants, such as dodecylsulfonate, and other species that are difficult to elute by conventional IC. Zipax SAX® has been used with success as an anion separator column.[331]

The ion chromatography has been widely used in a variety of applications. Sulfate is still the most frequently analyzed ion, but the number of ions and types of sample analyzable by IC has increased dramatically in the last few years. IC has become a common tool in air and water pollution analysis and its usefulness is just beginning to be recognized in food analysis and biomedical research. The development and proliferation of ion chromatographic methods should continue for many years.

NOTES

1. F. C. Smith, Jr. and R. C. Chang, "Ion Chromatography," *CRC Crit. Rev. Anal. Chem.* **9**(3), 197–217 (1980).

2. R. Bogoczek and G. Miemus, "Ion Chromatography—Another Ion Exchange Analytical Technique," *Przem. Chem.,* **59**(9), 471–474 (1980) (in Polish).

3. K. Oikawa, "Ion Chromatography," *Bunseki,* **8,** 531–538 (1980) (in Japanese).

4. W. E. Rich and R. A. Wetzel, "Ion Chromatography: A New Technique for Ion Analysis," *Actual Chem.,* **6,** 51–57 (1980) (in French).

5. T. Miller, "On-Stream Ion Chromatography: An Aid to Energy Conservation," *ISA Trans.,* **18**(2), 59–64 (1979).

6. A. J. Lipski and C. J. Vairo, "Applications of Ion Chromatography in an Analytical Services Laboratory," *Can. Res.,* **13**(1), 45–48 (1980).

7. J. C. MacDonald, "Ion Chromatography," *Am. Lab,* **11**(1), 45–55 (1979).

8. R. Wetzel, "Ion Chromatography-Further Applications," *Environ. Sci. Technol.,* **13**(10), 1214–1217 (1979).

9. Y. Tsuchitani, "Trace Analysis by Ion Chromatography," *Bunseki,* **9,** 603–605 (1979) (in Japanese).

10. J. D. Mulik and E. Sawicki, "Ion Chromatography," *Environ. Sci. Technol.*, **13**(7), 804–809 (1979).

11. H. Viikamo, "Ion Chromatography," *Kem.-Kemi*, **6**(4), 190–191 (1979) (in Finnish).

12. K. H. Jansen, "Ion Chromatography, A New Method in Analytical Chemistry," *Labor Praxis*, **2**(3), 30–34 (1978) (in German).

13. T. Nomura, "Principles and Applications of Ion Chromatography," *Kaguku to Kogyo*, **52**(12), 448–452 (1978) (in Japanese).

14. K. Oikawa, "Application of Ion Chromatography in Environmental Analysis," *PPM*, **9**(7), 52–61 (1978) (in Japanese).

15. A. J. Muir, "Analysis of Inorganic Ions by Ion Chromatography," *Sci. Technol.* (Surrey Hill, Aust.), **16**(2), 19–23 (1978).

16. K. H. Jansen, "Chromatography: Use in the Inorganic Analysis by Ion Chromatography," *GIT Fachz. Lab.*, **22**(12), 1062, 1064–1066, 1069–1071 (1978) (in German).

17. W. E. Rich, J. A. Tillotson, and R. C. Chang, "Ion Chromatography: An Analytical Perspective," in E. Sawicki, J. D. Mulik, and E. Wittgenstein, Eds., *Ion Chromatographic Analysis of Environmental Pollutants*, Vol. 1, Ann Arbor Science, Ann Arbor, Michigan, 1978, pp. 185–196.

18. H. Small, "An Introduction to Ion Chromatography," in E. Sawicki, J. D. Mulik, and E. Wittgenstein, Eds., *Ion Chromatographic Analysis of Environmental Pollutants*, Vol. 1, Ann Arbor Science, Ann Arbor, Michigan, (1978), pp. 11–21.

19. E. Sawicki, "Potential of Ion Chromatography for Environmental Studies," in E. Sawicki, J. D. Mulik, and E. Wittgenstein, Eds., *Ion Chromatographic Analysis of Environmental Pollutants*, Vol. 1, Ann Arbor Science, Ann Arbor, Michigan, 1978, pp. 1–9.

20. W. E. Rich, "Ion Chromatography," *Instrum. Technol.*, **24**(8), 47–51 (1978).

21. W. E. Rich, "Ion Chromatography: A New Technique for Automated Analysis of Ions in Solution," *Anal. Instrum.*, **15**, 113–117 (1977).

22. T. H. Maugh II, "IC Versatility Promotes Competition," *Science*, **208**(4), 164–165 (1980).

23. A. Jardy and R. Rosset, "Coupling Ion Exchange and Conductometric Detection: Ion Chromatography," *Analusis*, **7**(6), 259–267 (1979) (in French).

24. Dionex Auto Ion Application Note No. 1, "Analysis of Ambient Aerosol Filter Extracts" (1979).

25. Dionex Application Note No. 2R, "Analysis of Nitrate and Sulfate Collected on Air Filters" (1978).

26. L. D. Hansen, B. E. Richter, D. K. Rollins, J. D. Lamb, and D. J. Eatough, "Determination of Arsenic and Sulfur Species in Environmental Samples by Ion Chromatography," *Anal. Chem.*, **51**(6), 633–637 (1979).

27. J. D. Mulik, R. Puckett, E. Sawicki, and D. Williams, "Ion Chromatographic Analysis of Sulfate and Nitrate in Ambient Aerosols," *Anal. Lett.*, **9**(7), 653–663 (1976).

28. W. E. Rich and R. A. Wetzel, "Ion Chromatographic Analysis of Trace Ions in Environmental Samples," in D. Schuetzle, Ed., *Monitoring Toxic Substances, Symposium on Monitoring Toxic Substances of the 174th Meeting of ACS*, Chicago, Illinois, August 1977, American Chemical Society, Washington, D.C., 1979.

29. J. D. Mulik, R. Puckett, E. Sawicki, and D. Williams, "Ion Chromatography: A New Technique for Assay of Sulfate and Nitrate in Ambient Aerosols," *National Bureau of Standards Special Publication*, **464**, 603–607, (1977).

30. K. Anlauf, L. A. Barrie, A. A. Wiebe, and P. Fellin, "Major Ion Concentration in Atmospheric Particulates Sampled in Eastern Canada," *Can. Res.*, 49–53 (February 1980).

31. R. K. Stevens, T. G. Dzubay, G. Russwurm, and D. Rickel, "Sampling and Analysis of Atmospheric Sulfate and Related Species," *Atmos. Environ.*, (12), 55–68 (1978).

32. B. R. Appel and W. J. Wehrmeister, "An Evaluation of Sulfate Analysis of Atmospheric Samples by Ion Chromatography," in J. D. Mulik and E. Sawicki, Eds., *Ion Chromatographic Analysis of Environmental Pollutants*, Ann Arbor Science, Ann Arbor, Michigan, 1979, pp. 223–233.

33. F. E. Butler, R. H. Jungers, L. F. Porter, A. E. Riley, and F. J. Toth, "Analysis of Air Particulates by Ion Chromatography: Comparison With Accepted Methods," in E. Sawicki, J. D. Mulik, and E. Wittgenstein, Eds., *Ion Chromatographic Analysis of Environmental Pollutants*, Vol. 1, Ann Arbor Science, Ann Arbor, Michigan, 1978, pp. 65–76.

34. P. K. Mueller, B. V. Mendoza, J. C. Collins, and E. S. Wilgus, "Applications of Ion Chromatography to the Analysis of Anions Extracted from Airborne Particulate Matter," in E. Sawicki, J. D. Mulik, and E. Wittgenstein, Eds., *Ion Chromatographic Analysis of Environmental Pollutants*, Vol. 1, Ann Arbor Science, Ann Arbor, Michigan, 1978, pp. 77–86.

35. K. K. Fung, S. L. Heisler, A. Price, B. V. Neusca, and P. K. Mueller, "Comparison of Ion Chromatography and Automated Wet Chemical Methods for Analysis of Sulfate and Nitrate in Ambient Particulate Filter Samples," in J. D. Mulik and E. Sawicki, Eds., *Ion Chromatographic Analysis of Environmental Pollutants,* Vol. 2, Ann Arbor Sciences, Ann Arbor, Michigan, 1979, pp. 203–209.

36. D. A. Otterson, "Ion Chromatographic Determination of Anions Collected on Filters at Altitudes Between 9.6 and 13.7 Kilometers," in J. D. Mulik and E. Sawicki, Eds., *Ion Chromatographic Analysis of Environmental Pollutants,* Vol. 2, Ann Arbor Science, Ann Arbor, Michigan, 1979, pp. 87–98. See also *NASA Tech. Memo.,* NASA-TM-X-73642, 1977, 16 p.

37. D. A. Otterson, "Ion Chromatographic Determination of Microgram Quantities of Anions Collected on Cellulose Fiber Filters," presented at the Pittsburgh Conference on Analytical Chemistry and Applied Spectroscopy, Symposium on Ion Exchange and Ion Chromatography, Cleveland, Ohio, March 1978.

38. J. D. Mulik and E. Sawicki, "Ion Chromatography," *Environ. Sci. Technol.,* **13**(7), 804–809 (1979).

39. J. Lathouse and R. W. Coutant, "Practical Experience on the Use of Ion Chromatography for the Determination of Anions in Filter Catch Samples," in E. Sawicki, J. D. Mulik, and E. Wittgenstein, Eds., *Ion Chromatographic Analysis of Environmental Pollutants,* Vol. 1, Ann Arbor Science, Ann Arbor, Michigan, 1978, pp. 53–64.

40. L. M. Reisinger and T. L. Crawford, "Sulfur Measurements—Southeastern United States," presented at the 73rd Meeting of the Air Pollution Control Association, Montreal, Quebec, June 1980.

41. P. J. Galvin, P. J. Samson, P. E. Coffey, and D. Romano, "Transport of Sulfate in New York State," *Environ. Sci. Technol.,* **12**(5), 500–504 (1978).

42. D. F. Leahy, M. F. Phillips, R. W. Garber, and R. L. Tanner, "Filter Material for Sampling of Ambient Aerosols," *Anal. Chem.,* **52**(11), 1179–1180 (1980).

43. B. R. Appel, E. M. Hoffer, W. Wehrmeister, M. Haik, and J. J. Wesolowski, "Improvement and Evaluation of Methods for Sulfate Analysis. Parts 1 and 2," *PB-298,* 148 pages for Part I and EPA-R-805447-1 Final Report; 106 pages for Part II.

44. C. J. Jackson, D. Brussau, and C. Neuberger, "Ion Chromatographic Analysis of Gaseous and Particulate Fluoride Emissions from Brickworks and Aluminum Smelters," presented at the Rocky

Mountain Conference, Symposium on Ion Chromatography, Denver, Colorado, August 1981.

45. Dionex Application Note No. 2, "Analysis of Ambient Aerosol Filter Extracts" (1976).

46. J. D. Mulik, R. Puckett, E. Sawicki and O. Williams, "Ion Chromatographic Analysis of Sulfate and Nitrate in Ambient Aerosols," *Anal. Lett.* **9**(7), 653–663 (1976).

47. J. D. Mulik, E. Estes, and E. Sawicki, "Ion Chromatography Analysis of Ammonium Ion in Ambient Aerosols," in E. Sawicki, J. D. Mulik, and E. Wittgenstein, Eds., *Ion Chromatographic Analysis of Environmental Pollutants,* Vol. 1, Ann Arbor Science, Ann Arbor, Michigan, 1978, pp. 41–51.

48. J. D. Mulik, R. Puckett, E. Sawicki, and D. Williams, "Ion Chromatography Analysis of Sulfate and Nitrate in Ambient Aerosols," *Anal. Lett.,* **9**(7), 653–663 (1976).

49. D. W. Mason and H. C. Miller, "Measurement of Ambient Sulfuric Acid Aerosol with Analysis by Ion Chromatography," in J. D. Mulik and E. Sawicki, *Ion Chromatographic Analysis of Environmental Pollutants,* Vol. 2, Ann Arbor Science, Ann Arbor, Michigan, 1979, pp. 193–201.

50. C. W. Lewis and E. S. Macias, "Composition of Size-Fractionated Aerosol in Charleston, West Virginia," *Atmos. Environ.,* **14**, 185–194, (1980).

51. R. Chul-Un, W. H. Chan, M. A. Lusis, and R. Vet, "Analysis of Precipitation Measurements in the Vicinity of a Nickel Smelter," presented at the 73rd Air Pollution Control Association Meeting, Montreal, Quebec, June 1980.

52. J. Slanina, W. A. Ingerak, J. E. Ordelman, P. Borst, and F. P. Bakker, "Automation of Ion Chromatography and Adaptation for Rainwater Analysis," in J. D. Mulik and E. Sawicki, Eds., *Ion Chromatographic Analysis of Environmental Pollutants,* Vol. 2, Ann Arbor Science, Ann Arbor, Michigan, 1979, pp. 305–318.

53. H. M. Liljestrand and J. J. Morgan, "Chemical Composition of Acid Precipitation in Pasadena, California," *Environ. Sci. Technol.,* **12**(12), 1271–1273 (1978).

54. D. Pickerell, T. Hook, T. Dolzine, and J. K. Robertson, "Intensity-Weighted Sequential Sampling of Precipitation: A Technique for Monitoring Changes in Storm Chemistry During a Storm," in J. D. Mulik and E. Sawicki, Eds., *Ion Chromatographic Analysis of Environmental Pollutants,* Vol. 2, Ann Arbor Science, Ann Arbor, Michigan, 1979, pp. 289–294.

55. D. C. Bogen and S. J. Nagourney, "Ion Chromatographic Analysis of Cations at Baseline Precipitation Stations," in J. D. Mulik and E. Sawicki, Eds., *Ion Chromatographic Analysis of Environmental Pollutants,* Vol. 2, Ann Arbor Science, Ann Arbor, Michigan, 1979, pp. 319–328.

56. S. Y. Tyree, Jr., J. M. Stouffer, and M. Bollinger, "Ion Chromatographic Analysis of Simulated Rain Water," in J. D. Mulik and E. Sawicki, Eds., *Ion Chromatographic Analysis of Environmental Pollutants,* Vol. 2, Ann Arbor Science, Ann Arbor, Michigan, 1979, pp. 295–304.

57. P. J. Galvin and J. A. Cline, "Measurement of Anions in the Snow Cover of the Adirondack Mountains," *Atmos. Environ.,* **12**(5), 1163–1167 (1978).

58. J. Rothert, "Use of Ion Chromatography for Analysis of MAP3S Precipitation Samples," *DOE Report No.* CONF-800837-2.

59. J. Slanina and F. P. Bakker, "Development of High Speed Ion Chromatography," presented at the Rocky Mountain Conference, Symposium on Ion Chromatography, Denver, Colorado, August 1981.

60. J. Slanina, W. A. Ingerak, J. E. Ordelman, P. Borst, and F. P. Bakker, "Automation of Ion Chromatography and Adaptation for Rain Water Analysis," in J. D. Mulik and E. Sawicki, Eds., *Ion Chromatographic Analysis of Environmental Pollutants,* Vol. 2, Ann Arbor Science, Ann Arbor, Michigan, 1979, pp. 305–318.

61. "Procedures Manual Level I Environmental Assessment," EPA Report, EPA-600/7-78-201.

62. A. Fitchett, Dionex Corporation, Sunnyvale, California, private communication.

63. P. K. Dasgupta, K. DeCesare, and J. C. Ullrey, "Determination of Atmospheric Sulfur Dioxide without Tetrachloromercurate (II) and the Mechanism of the Schiff Reaction," *Anal. Chem.* **52**(12), 1912–1922 (1980).

64. J. D. Mulik, G. Todd, E. Estes, R. Puckett, E. Sawicki, and D. Williams, "Ion Chromatographic Determination of Atmospheric Sulfur Dioxide," in J. D. Mulik and E. Sawicki, Eds., *Ion Chromatographic Analysis of Environmental Pollutants,* Vol. 1, Ann Arbor Science, Ann Arbor, Michigan, 1978, pp. 23–40.

65. B. Dellinger, G. Grotecloss, C. R. Fortune, J. L. Cheney, and J. B. Homolya, "Sulfur Dioxide Oxidation and Plume Formation at Cement Kilns," *Environ. Sci. Technol.,* **14**(10), 1244–1249 (1980).

66. D. L. Smith, W. S. Kim, and R. E. Kupel, "Determination of Sulfur Dioxide by Adsorption on a Solid Sorbent Followed by Ion Chromatography Analysis," *Am. Indus. Hyg. Assoc. J.*, **41**(7), 485–488 (1980).

67. C. D. Frazier, "Evaluation of Ion Chromatography as an Equivalent Method for Ambient Sulfur Dioxide Analysis," in J. D. Mulik and E. Sawicki, Eds., *Ion Chromatographic Analysis of Environmental Pollutants*, Vol. 2, Ann Arbor Science, Ann Arbor, Michigan, 1979, pp. 211–221.

68. C. J. Holcombe and J. C. Terry, "Application of Ion Chromatography for the Analysis of Lime/Limestone Based Sulfur Dioxide Scrubber Solutions," *Proc. Ann. Meet. Air Pollut. Control Assoc.*, **71**(5), 78/71.6 (1978) 14 p.

69. C. D. Frazier, "Evaluation of Ion Chromatography as an Equivalent Method for Ambient Sulfur Dioxide Analysis," in J. D. Mulik and E. Sawicki, Eds., *Ion Chromatographic Analysis of Environmental Pollutants*, Vol. 2, Ann Arbor Science, Ann Arbor, Michigan, 1979, pp. 211–221.

70. J. D. Mulik, G. Todd, E. Estes, R. Puckett, E. Sawicki, and D. Williams, "Ion Chromatographic Determination of Atmospheric Sulfur Dioxide," in J. D. Mulik and E. Sawicki, Eds., *Ion Chromatographic Analysis of Environmental Pollutants*, Vol. 1, Ann Arbor Science, Ann Arbor, Michigan, 1978, pp. 23–40.

71. P. K. Dasgupta, K. DeCesare, and J. C. Ullery, "Determination of Atmospheric Sulfur Dioxide Without Tetrachloromercurate (II) and the Mechanism of the Schiff Reaction," *Anal. Chem.*, **52**(12), 1912–1922 (1980).

72. F. C. Smith, Jr., Microsensor Technology Inc., Fremont, California, unpublished work.

73. D. V. Vinjamoori and C. Ling, "Personal Monitoring Method For Nitrogen Dioxide and Sulfur Dioxide with Solid Sorbent Sampling and Ion Chromatographic Determinations," *Anal. Chem.*, **53**(11), 1689–1690 (1981).

74. J. H. Dempsey, P. Curse, and K. Yates, "Analysis of Anions in Flue Gas Desulfurization Systems by Ion Chromatography," in E. Sawicki, J. D. Mulik, and E. Wittgenstein, Eds., *Ion Chromatographic Analysis of Environmental Pollutants*, Vol. 1, Ann Arbor Science, Ann Arbor, Michigan, 1978, pp. 89–97.

75. R. Steiber and R. M. Statnick, "Application of Ion Chromatography to Stationary Source and Control Device Evaluation Studies," in E. Sawicki, J. D. Mulik and E. Wittgenstein, Eds., *Ion*

Chromatographic Analysis of Environmental Pollutants, Vol. 1, Ann Arbor Science, Ann Arbor, Michigan, 1978, pp. 141–148.

76. L. J. Holcombe and J. C. Terry, "Application of Ion Chromatography for the Analysis of Lime/Limestone Based Sulfur Dioxide Scrubber Solutions," *Proc. Ann. Meet. Air Pollut. Control Assoc.,* **71**(5), 78/71.6, (1978) (14 p.).

77. R. Steiber and R. Merrill, "Application of Ion Chromatography to the Analysis of Source Assessment Samples," in J. D. Mulik and E. Sawicki, Eds., *Ion Chromatographic Analysis of Environmental Pollutants,* Vol. 2, Ann Arbor Science, Ann Arbor, Michigan, 1979, pp. 99–113.

78. T. R. Acciani and R. F. Maddalone, "Chemical Analysis of Wet Scrubbers Utilizing Ion Chromatography," EPA Contract 68-02-2165, 1979, 63 p.

79. R. F. Maddalone, L. L. Scinto, and M. M. Yamada, "Sampling and Analysis of Reduced and Oxidized Species in Process Streams," EPA Contract 600/2079-201, November 1979, 229 p.

80. J. R. Donnelly, D. C. Shepley, T. M. Martin, and A. H. Abdulsattar, "Laboratory Procedures: Analysis of Sodium-Based Dual Alkali Process Streams," EPA Contract 68-02-2634, March 1980, 162 p.

81. F. B. Meserole, D. L. Lewis, A. W. Nichols, and G. Rochelle, "Adipic Acid Degradation Mechanism in Aqueous FGD (Flue Gas Desulfurization) Systems," EPA Contract 68-02-2608, September 1979, 93 p.

82. Dionex Applications Note 12, "Analysis of Ions in Flue Gas Scrubber Solutions."

83. D. V. Vinjamoori and C. Ling, "Personal Monitoring Method For Nitrogen Dioxide and Sulfur Dioxide with Solid Sorbent Sampling and Ion Chromatographic Determinations," *Anal. Chem.,* **53**(11), 1689–1690 (1981).

84. D. O. Saunders, Union Oil, Brea, California, private communication.

85. J. C. Terry and E. E. Ellsworth, "IC Analysis of Formaldehyde Stabilized Sulfite Solutions," presented at the Rocky Mountain Conference, Denver, Colorado, August 1980.

86. L. D. Hansen, B. E. Richter, D. K. Rollins, J. D. Lamb, and D. J. Eatough, "Determination of Arsenic and Sulfur Species in Environmental Samples by Ion Chromatography," *Anal. Chem.,* **51**(6), 633–637 (1979).

87. R. F. Maddalone, L. L. Scinto, and M. M. Yamada, "Sampling and Analysis of Reduced and Oxidized Species in Process Streams," Report for EPA Contract 68-02-2165, November 1979, 229 p.

88. "Analysis of Ions in Flue Gas Scrubber Solutions," Dionex Corporation, Sunnyvale, California, Application Note 12.

89. J. R. Donnelly, D. C. Shepley, T. M. Martin, and A. H. Abdulsatter, "Laboratory Procedures: Analysis of Sodium-Based Dual-Alkali Process Streams," Report for EPA Contract 68-02-2634, March 1980, 162 p.

90. F. B. Meserole, D. L. Lewis, A. N. Nichols, and G. Rochelle, "Adipic Acid Degradation Mechanism in Aqueous FGD (Flue Gas Desulfurization) Systems," Report for EPA Contract 68-02-2608, September 1979, 93 p.

91. J. C. Terry, E. E. Ellsworth, and D. L. Utley, "ICE Analysis of Adipic Acid Degradation Products in Limestone FGD Scrubber Solutions," presented at the Rocky Mountain Conference, Symposium on Ion Chromatography, Denver, Colorado, August 1981.

92. R. B. Zweidinger, S. B. Tejada, J. E. Sigsby, Jr., and R. L. Bradow, "Application of Ion Chromatography to the Analysis of Ammonia and Alkylamines in Automobile Exhaust," in E. Sawicki, J. D. Mulik, and E. Wittgenstein, Eds., *Ion Chromatographic Analysis of Environmental Pollutants*, Vol. 1, Ann Arbor Science, Ann Arbor, Michigan 1978, pp. 125–139.

93. J. M. Clingenpeel and D. E. Seizinger, "Computerized Analysis of Sulfates and Ammonia in Diesel Samples Using Ion Chromatography," presented at the Rocky Mountain Conference, Denver, Colorado, August 1980.

94. T. J. Truex, W. R. Pierson, D. E. McKee, M. Shelef, and R. E. Baker , "Effects of Barium Fuel Additive and Fuel Sulfur Level on Diesel Particulate Emmissions," *Environ. Sci. Technol.*, **14**(9), 1121 (1980).

95. S. B. Tejada, R. B. Zweidinger, J. E. Sigsby, Jr., and R. L. Bradow, "Modification of an Ion Chromatograph for Automated Routine Analysis: Application to Mobile Source Emission," in E. Sawicki, J. D. Mulik, and E. Wittgenstein, Eds., *Ion Chromatographic Analysis of Environmental Pollutants,* Vol. 1, Ann Arbor Science, Ann Arbor, Michigan, 1978, pp. 111–124.

96. N. P. Barkley, G. L. Contner, and M. Malanchuk, "Simultaneous Analysis of Anions and Cations in Diesel Exhaust Using Ion Chromatography," in J. D. Mulik and E. Sawicki, Eds., *Ion Chroma-*

tographic Analysis of Environmental Pollutants, Vol. 2, Ann Arbor Science, Ann Arbor, Michigan, 1979, pp. 115–128.

97. I. Bodek and K. T. Menzies, "Analysis of Organic Acids from Diesel Exhaust in Mine Air," presented at the 2nd Symposium on Process Measurement for Environmental Assessment, February 1980.

98. D. W. Mason and H. C. Miller, "Measurement of Ambient Sulfuric Acid Aerosol with Analysis by Ion Chromatography," in J. D. Mulik and E. Sawicki, Eds., *Ion Chromatographic Analysis of Environmental Pollutants,* Vol. 2, Ann Arbor Science, Ann Arbor, Michigan, 1979, pp. 193–202.

99. D. A. Otterson, "Ion Chromatographic Determination of Anions Collected on Filters at Altitudes Between 9.6 and 13.7 Kilometers," NASA Report NASA-TM-X-73642, E-9143, 1977, 16 p. See also E. Sawicki, J. D. Mulik, and E. Wittgenstein, Eds., *Ion Chromatographic Analysis of Environmental Pollutants,* Vol. 1, Ann Arbor Science, Ann Arbor, Michigan, 1978, pp. 87–98.

100. B. Hubert, Colorado State University, Colorado Springs, Colorado, private communication.

101. S. Wilson, C. Gent, and T. Hinkley, "Analysis of Mount St. Helens Samples Using Ion Chromatography," presented at the Rocky Mountain Conference, Symposium on Ion Chromatography, Denver, Colorado, August 1981.

102. C. J. Jackson, D. Bussan, and C. Neuberger, "Ion Chromatographic Analysis of Gaseous and Particulate Fluoride Emissions from Brickworks and Aluminum Smelters," presented at the Rocky Mountain Conference, Symposium on Ion Chromatography, Denver, Colorado, August 1981.

103. J. H. Lowry, K. Wang, and R. C. Ross, "Ion Chromatographic Analysis of Selected Anions Associated with Copper Smelting Operations," presented at the Rocky Mountain Conference, Denver, Colorado, August 1980.

104. R. Chul-un, W. H. Chan, M. A. Lusis, and R. Vet, "Analysis of Precipitation Measurements in the Vicinity of a Nickel Smelter," presented at 73rd Annual Meeting of the Air Pollution Control Association, Montreal, Quebec, June 1980.

105. H. Matusiewicz and D. F. S. Natusch, "Ion Chromatographic Determination of Soluable Anions Present in Coal Fly Ash Leachates," *Int. J. Environ. Anal. Chem.,* 8(3), 227–233 (1980).

106. B. F. Jones III and K. Schwitzgebel, "Potential Groundwater Contamination Resulting from the Disposal of Flue Gas Cleaning Wastes, Fly Ash, and SO_2 Scrubber Sludge," presented at 71st Meeting of the Air Pollution Control Association, Houston, Texas, June 1978.

107. I. Bodek and R. H. Smith, "Determination of Ammonium Sulfamate in Air Using Ion Chromatography," *Am. Indust. Hyg. Assoc. J.,* **41**(8), 603–607 (1980).

108. L. C. Westwood and E. L. Stokes, "Determination of Azide in Environmental Samples by Ion Chromatography," in J. D. Mulik and E. Sawicki, Eds., *Ion Chromatographic Analysis of Environmental Pollutants,* Vol. 2, Ann Arbor Science, Ann Arbor, Michigan, 1979, pp. 141–156.

109. L. D. Hansen, J. F. Ryder, N. F. Mangelson, M. W. Hill, K. J. Faucet, and D. J. Eatough, *Anal. Chem.,* **52**, 821–824 (1980).

110. R. W. Shaw, R. K. Stevens, and W. J. Courtney; D. A. Hegg, P. V. Hobbs, L. D. Hansen, M. F. Mangelson, and D. J. Eatough, *Anal. Chem.,* **52**, 2217–2219 (1980).

111. E. Sawicki, J. D. Mulik, and E. Wittgenstein, Eds., *Ion Chromatographic Analysis of Environmental Pollutants,* Vol. 1, Ann Arbor Science, Ann Arbor, Michigan, 1978, 210 p.

112. J. D. Mulik and E. Sawicki, Eds., *Ion Chromatographic Analysis of Environmental Pollutants,* Volume 2, Ann Arbor Science, Ann Arbor, Michigan, 1979, 435 p.

113. R. Steiber and R. Merrill, "Application of Ion Chromatography to the Analysis of Source Assessment Samples," in J. D. Mulik and E. Sawicki, Eds., *Ion Chromatographic Analysis of Environmental Pollutants,* Vol. 2, Ann Arbor Science, Ann Arbor, Michigan, 1979, pp. 99–114.

114. R. Steiber and R. Merrill, "Determination of Arsenic as the Oxidate by Ion Exchange," *Anal. Lett.,* **12**(A3), 273–278 (1979).

115. "Determination of As^{+3} and As^{+5}," Dionex Corporation, Sunnyvale, California, Application Note 22.

116. L. D. Hansen, B. E. Richter, D. K. Rollins, J. D. Lamb, and D. J. Eatough, "Determination of Arsenic and Sulfur Species in Environmental Samples by Ion Chromatography," *Anal. Chem.,* **51**(6), 633–637 (1979).

117. G. R. Ricci, L. S. Shepard, G. Colovos, and N. E. Hester, "Ion Chromatography with Atomic Absorption Spectrometric Detection for Determination of Organic and Inorganic Arsenic Species," *Anal. Chem.,* **53**(4), 610–613 (1981).

118. L. D. Hansen, S. F. Ryder, N. F. Mangelson, N. W. Hill, K. J. Faucet, and D. J. Eatough, *Anal. Chem.*, **52**, 821 (1980).

119. R. W. Shaw, R. K. Stevens, W. J. Courtney, D. A. Hegg, P. V. Hobbs, L. D. Hansen, N. F. Mangelson, and D. J. Eatough, *Anal. Chem.*, **52**, 2217–2219 (1980).

120. J. M. Lorraine, C. R. Fortune, and B. Dellinger, "Sampling and Ion Chromatographic Determination of Formaldehyde and Acetaldehyde," *Anal. Chem.*, **53**(8), 1302–1305 (1981).

121. W. S. Kim, C. L. Geraci, and R. E. Kupel, "Solid Sorbent Tube Sampling and Ion Chromatographic Analyses of Formaldehyde," *Am. Indus. Hyg. Assoc. J.*, **41**(5), 334–339 (1980).

122. D. L. Haynes, "Collection of Formic Acid Vapor and Analysis by Ion Chromatography," in J. D. Mulik and E. Sawicki, Eds., *Ion Chromatographic Analysis of Environmental Pollutants,* Vol. 2, Ann Arbor Science, Ann Arbor, Michigan, 1979, pp. 157–169.

123. W. S. Kim, C. L. Geraci, Jr., and R. E. Kupel, "Sampling and Analysis of Formaldehyde in Industrial Atmospheres," in J. D. Mulik and E. Sawicki, Eds., *Ion Chromatographic Analysis of Environmental Pollutants,* Vol. 2, Ann Arbor Science, Ann Arbor, Michigan, 1979, pp. 171–184.

124. K. K. Beasley, C. E. Hoffman, M. L. Rueppel, and J. W. Worley, "Sampling of Formaldehyde in Air with Coated Solid Sorbent and Determination by High Performance Liquid Chromatography," *Anal. Chem.*, **52**, 1110–1114 (1980).

125. "Determination of Formaldehyde as Formate Ion," Dionex Corporation, Sunnyvale, California, Application Note 24.

126. D. L. Haynes, "Collection of Formic Acid Vapor and Analysis by Ion Chromatography," in J. D. Mulik and E. Sawicki, Eds., *Ion Chromatography Analysis of Environmental Pollutants,* Vol. 2, Ann Arbor Science, Ann Arbor, Michigan, 1979, pp. 157–169.

127. C. J. Jackson and D. Bussan, "Ion Chromatography Analysis of Formaldehyde: An Evaluation of Proprietary Solid-Absorber Sampling Tubes Under Laboratory and Field Conditions," presented at the Rocky Mountain Conference, Symposium on Ion Chromatography, Denver, Colorado, August 1981.

128. T. W. Dolzine, G. G. Exposito, and R. J. Gaffney, "Determination of Hydrogen Cyanide in Environmental Samples by Ion Chromatography," presented at the Rocky Mountain Conference Symposium on Ion Chromatography, Denver, Colorado, August 1981.

129. P. R. McCullough and J. W. Worley, "Sampling of Chloracetyl Chloride in Air on Solid Support and Determination by Ion Chromatography," *Anal. Chem.*, **51**(8), 1120–1122 (1979).

130. "Determination of Monochloroacetyl Chloride," Dionex Corporation, Sunnyvale, California, Application Note 23.

131. L. C. Westwood and E. L. Stokes, "Determination of Azide in Environmental Samples by Ion Chromatography," *Anal. Lett.*, June 1978.

132. L. C. Westwood and E. L. Stokes, "Determination of Azide in Environmental Samples by Ion Chromatography," in J. D. Mulik and E. Sawicki, Eds., *Ion Chromatographic Analysis of Environmental Pollutants*, Vol. 2, Ann Arbor Science, Ann Arbor, Michigan, 1979, pp. 141–156.

133. "Azide Determination in Complex Matrices," Dionex Corporation, Sunnyvale, California, Application Note 14.

134. F. C. Smith, Jr., Microsensor Technology Inc., Fremont, California, unpublished work.

135. J. Septon, OSHA, Salt Lake City, Utah, private communication.

136. I. Kifune and K. Oikawa, "Determination of Trace Amounts of Ammonia and Lower Amines in the Atmosphere by Ion Chromatography," *Bunseki Kagaku*, **28**(10), 587–598 (1979) (in Japanese).

137. I. Kifune and K. Oikawa, "Determination of Trace Ammonia and Amines in Environmental Air Using Ion Chromatography," *Nigata Rikagaku*, **5**, 9–14 (1979) (in Japanese).

138. W. S. Kim, J. D. McGlothlin, and R. E. Kupel, "Sampling and Analysis of Iodine in Industrial Atmosphere," *Am. Indus. Hyg. Assoc. J.*, **42**(3), 187–190 (1981).

139. R. D. Holm and S. A. Barksdale, "Analysis of Anions in Combustion Products," in E. Sawicki, J. D. Mulik, and E. Wittgenstein, Eds., *Ion Chromatographic Analysis of Environmental Pollutants*, Vol. 1, Ann Arbor Science, Ann Arbor, Michigan, 1978, pp. 99–110.

140. L. C. Speitel, J. C. Spurgeon, and R. A. Filipczak, "Ion Chromatographic Analysis of Thermal Decomposition Products of Aircraft Interior Materials," in J. D. Mulik and E. Sawicki, Eds., *Ion Chromatographic Analysis of Environmental Pollutants*, Vol. 2, Ann Arbor Science, Ann Arbor, Michigan, 1979, pp. 75–88.

141. "Determination of Thermal Decomposition Products of Polymeric Materials," Dionex Corporation, Sunnyvale, California, Application Note 13.

142. P. White, "Ion Chromatography: Evaluation of an Ion Chromatograph," internal report of the Metropolitan Police, Forensic Science Laboratory, London, England (1979).

143. F. C. Smith, Jr., Microsensor Technology Inc., Fremont, California, and J. E. Menear, Dionex Corporation, Ft. Worth, Texas, unpublished work.

144. D. J. Rutter, R. C. Buechele, T. L. Randolph, and E. C. Bender, "Identification of Explosives and Explosive Residues by Ion Chromatography," presented at the Rocky Mountain Conference, Symposium on Ion Chromatography, Denver, Colorado, August 1981.

145. M. Lindgren, "Ion Chromatography in Water Analysis," *Vatten*, **36**, 249–264 (1980) (in Swedish).

146. C. Pohlandt, "Determination of Anions by Ion Chromatography," *S. Afr. J. Chem.*, **33**(3), 89–91 (1980).

147. D. Kasiske and M. Sonnborn, "Analysis of Anionic Constituents in Natural Waters by Ion Chromatography," *Labor Praxis*, **4**(4), 76–83 (1980) (in German).

148. M. J. Fishman and G. Pyen, "Determination of Selected Anions in Water by Ion Chromatography," USGS Reports USGS/WRI-79-101 and USGS/WRD/WRI-79/083, 37 p.

149. C. Pohlandt, "The Separation and Determination of Anions by Ion Chromatography," *Rep.-Natl. Inst. Metall. (S. Afr.)*, **2044**, 1980, 21 p.

150. B. W. Smee, G. E. M. Hall, and D. J. Koop, "Analysis of Fluoride, Chloride, Nitrate, and Sulfate in Natural Waters Using Ion Chromatography," *J. Geochem. Explor.*, **10**(3), 245–258 (1978).

151. T. S. Long and A. L. Reinsvold, "Application of Ion Chromatography to the Analysis of Aqueous Solutions," *Jt. Conf. Sens. Environ. Pollut.*, (1978), pp. 624–629.

152. J. A. Rawa, "Application of Ion Chromatography to Analysis of Industrial Process Waters," in E. Sawicki, J. D. Mulik, and E. Wittgenstein, Eds., *Ion Chromatographic Analysis of Environmental Pollutants*, Vol. 1, Ann Arbor Science, Ann Arbor, Michigan, 1978, pp. 245–269.

153. J. A. Rawa and E. L. Henn, "Characterization of Industrial Process Waters and Water-formed Deposits by Ion Chromatography," *Proc. Int. Water Conf. Eng. Soc. West. Pa.*, **40**, 213–219 (1979).

154. W. E. Rich, "Ion Chromatography: A New Technique for the Automated Analysis of Ions in Solution," in *Analysis Instrumentation*, Vol. 15, Instrument Society of America, Pittsburgh, Pennsylvania, 1977, pp. 113–117.

155. R. Chang, "Ion Chromatography: A New Technique for Water Analysis," presented at the Pittsburgh Conference on Analytical

Chemistry and Applied Spectroscopy, Cleveland, Ohio, March 1977.

156. "Analysis of Industrial Wastestreams," Dionex Corporation, Sunnyvale, California, Application Note 11.

157. "Anions in Wastewater," Wescan Corporation, Santa Clara, California, Ion Analysis Application No. 8.

158. W. H. Ficklin, "Separation and Analysis of Tungsten and Molybdenum in Natural Waters by Ion Chromatography," presented at the Rocky Mountain Conference, Denver, Colorado, August 1980.

159. W. E. Rich, Dionex Corporation, Sunnyvale, California, unpublished work.

160. M. J. Fishman and G. Pyen, "Determination of Selected Anions in Water by Ion Chromatography," USGS Reports USGS/WRI-79-101 and USGS/WRD/WRI-79/083, 37 p.

161. "Analysis of Micronutrients at Low Parts per Billion Levels," Dionex Corporation, Sunnyvale, California, Internal Applications Laboratory Report.

162. H. Small, T. S. Stevens, and W. C. Bauman, "Novel Ion Exchange Chromatographic Method Using Conductimetric Detection," *Anal. Chem.,* **47**(11), 1801–1809 (1975).

163. T. S. Stevens, Dow Chemical Company, Midland, Michigan, private communication.

164. J. A. Rawa and E. L. Henn, "Characterization of Industrial Process Wastes and Water-formed Deposits by Ion Chromatography," *Proc. Int. Water Conf. Eng. Soc. West. Pa.,* **40**, 213–219 (1979).

165. J. A. Rawa, "Impact of Ion Chromatography on the Water Treatment Industry," presented at the Rocky Mountain Conference, Symposium on Ion Chromatography, Denver, Colorado, August 1981.

166. R. A. Zweidinger, E. O. Estes, and L. W. Little, "Analysis of Monosodium Methanearsonate in Raw and Biologically Treated Combined Industrial–Municipal Wastes," presented at the Pittsburgh Conference on Analytical Chemistry and Applied Spectroscopy, Atlantic, City, New Jersey, March 1980.

167. G. S. Pyen and M. J. Fishman, "Determination of Anions in Pore Waters from Cores by Ion Chromatography," in J. D. Mulik and E. Sawicki, Eds., *Ion Chromatographic Analysis of Environmental Pollutants,* Vol. 2, Ann Arbor Science, Ann Arbor, Michigan, 1979, pp. 235–244.

168. T. Takamatsu, M. Kawashima, and M. Koyama, "Ion Chroma-

tographic Determination of Arsenite and Arsenate in Sediment Extract," *Bunseki Kagaku,* **28**(10), 596–601 (1979) (in Japanese).

169. C. J. Hill and R. P. Lash, "Evaluation of Ion Chromatography for Determination of Selected Ions in Geothermal Well Water," *Anal. Chem. Acta,* **108**, 405–409 (1979).

170. H. Itoh and Y. Shinbori, "Determination of Anions in Sea Water by Ion Chromatography," *Bunseki Kagaku,* **29**(4), 239–243 (1980).

171. D. D. Siemer, "Separation of Chloride and Bromide from Complex Matrices Prior to Ion Chromatographic Determination," *Anal. Chem.,* **52**(12), 1874–1877 (1980).

172. "Trace Sulfate and Phosphate in Brine," Dionex Corporation, Sunnyvale, California, Application Note No. 3.

173. R. M. Merrill and R. J. Kottenstette, "Ion Chromatographic Determination of Bromide and Nitrate in Geological Brines," presented at the Rocky Mountain Conference, Denver, Colorado, August 1980.

174. R. Chang, "Ion Chromatography, A New Technique for Water Analysis," presented at the Pittsburgh Conference on Analytical Chemistry and Applied Spectroscopy, Cleveland, Ohio, March 1977.

175. A. Fitchett, Dionex Corporation, Sunnyvale, California, private communication.

176. F. C. Smith, Jr., Microsensor Technology Inc., Fremont, California, unpublished work.

177. T. S. Stevens, V. T. Turkelson, and W. R. Albe, "Determination of Anions in Boiler Blow-Down Water with Ion Chromatography," *Anal. Chem.,* **49**(8), 1176–1178 (1977).

178. "Analysis of Industrial Wastestreams," Dionex Corporation, Sunnyvale, California, Application Note No. 11.

179. Wescan Corporation, Santa Clara, California, Ion Analysis Applications No. 8.

180. "Analyses of Oxalate in Industrial Liquors," Dionex Corporation, Sunnyvale, California, Applications Note No. 5.

181. "Anions in Wood Pulp Processing Liquor," Wescan Corporation, Santa Clara, California, Ion Analysis Application No. 10.

182. R. M. Felder, R. M. Kelly, J. K. Ferrell, and R. W. Rousseau, "How Clean Gas is Made from Coal," *Environ. Sci. Technol.,* **14**(6), 658–666 (1980).

183. K. M. McFadden and T. R. Garland, "Determination of Sulfur Species in Oil Shale Waste Waters by Ion Chromatography," Report on Contract AC06-76 RL01830, October 1980, 20 p.

184. R. Wetzel, F. C. Smith, Jr., and E. Cathers, "Industrial Applications of Ion Chromatography," *Indus. Res. Dev.*, February 1980.

185. G. H. Mansfield, "Ion Analysis in Chemical Plant Control and the Effect on Effluents," in J. D. Mulik and E. Sawicki, Eds., *Ion Chromatographic Analysis of Environmental Pollutants*, Vol. 2, Ann Arbor Sciences, Ann Arbor, Michigan, 1979, pp. 271–288.

186. T. Miller, "On-Stream Ion Chromatography: An Aid to Energy Conservation," *ISA Trans.*, **18**(2), 59–64 (1979).

187. D. Saunders and M. Winward, Union Oil Corporation, Brea, California, private communication.

188. "Determination of Perchlorate," Dionex Corporation, Sunnyvale, California, Application Note No. 18R.

189. M. J. McCormack, "Determination of Total Sulfur in Fuel Oils by Ion Chromatography," *Anal. Chem. Acta*, **121**, 233–238 (1980).

190. F. E. Butler, F. J. Toth, D. J. Driscoll, J. N. Hein, and R. H. Jungers, "Analysis of Fuels by Ion Chromatography: Comparison with ASTM Methods," in J. D. Mulik and E. Sawicki, Eds., *Ion Chromatographic Analysis of Environmental Pollutants*, Vol. 2, Ann Arbor Science, Ann Arbor, Michigan, 1979, pp. 185–192.

191. C. S. Mizisin, D. E. Kuivinen, and D. A. Otterson, "Ion Chromatographic Determination of Sulfur in Fuels," in J. D. Mulik and E. Sawicki, Eds., *Ion Chromatographic Analysis of Environmental Pollutants*, Vol. 2, Ann Arbor Science, Ann Arbor, Michigan, 1979, pp. 129–139.

192. R. Grabowski, Shell Oil Company, Houston, Texas, private communication.

193. W. J. Caveny and J. D. Childs, "Some Applications of Ion Chromatography to Oil Well Cement Research and Analysis," presented at the Rocky Mountain Conference, Symposium on Ion Chromatography, Denver, Colorado, August 1981.

194. W. Rich, F. Smith, Jr., L. McNeil, and T. Sidebottom, "Ion Exclusion Coupled to Ion Chromatography: Instrumentation and Application," in J. D. Mulik and E. Sawicki, Eds., *Ion Chromatographic Analysis of Environmental Pollutants*, Vol. 2, Ann Arbor Science, Ann Arbor, Michigan, 1979, pp. 17–29.

195. F. C. Smith, Jr., "Total Carbonate Species Determination by Ion Exclusion Chromatography," presented at Dionex Symposium on Ion Chromatography, Sunnyvale, California, June 1978.

196. F. C. Smith, Jr., Microsensor Technology Inc., Fremont, California, unpublished work.

197. A. Fitchett, Dionex Corporation, Sunnyvale, California, private communication.

198. R. Chang, General Electric, Mt. Vernon, Indiana, unpublished work.

199. A. Liao, Monsanto, Texas City, Texas, private communication.

200. V. Smith, Dionex Corporation, Sunnyvale, California, private communication.

201. J. P. Price, N. W. Stacy, and J. R. Hawkins, "Analysis of Fluoride in Nylon," presented at the Rocky Mountain Conference, Symposium on Ion Chromatography, Denver, Colorado, August 1981.

202. M. Williams, Dionex Corporation, New Jersey, private communication.

203. L. J. Schiff, S. G. Pleva, and E. W. Sarver, "Analysis of Phosphonic Acids by Ion Chromatography," in J. D. Mulik and E. Sawicki, Eds., *Ion Chromatographic Analysis of Environmental Pollutants,* Vol. 2, Ann Arbor Science, Ann Arbor, Michigan, 1979, pp. 329–344.

204. S. A. Bouyoucos and D. N. Armentrout, "Determination of Organophosphoric and Organophosphorothioic Acids in Aqueous Solutions by Ion Chromatography," *J. Chromatogr.,* **189**(1), 61–71 (1980).

205. W. E. Rich, Dionex Corporation, Sunnyvale, California, private communication.

206. F. C. Smith, Jr., Microsensor Technology Inc., Fremont, California, unpublished work.

207. E. L. Johnson, Dionex Corporation, Sunnyvale, California, private communication.

208. Dionex Training Course, 1977, p. IV-17.

209. R. K. Pinschmidt, Air Products, Fogelsville, Pennsylvania, private communication.

210. R. K. Pinschmidt and T. P. Katrinack, "Analysis of Chloride and Other Ions in Acetic Acid," in J. D. Mulik and E. Sawicki, Eds., *Ion Chromatographic Analysis of Environmental Pollutants,* Volume 2, Ann Arbor Science, Ann Arbor, Michigan, 1979, pp. 31–40.

211. M. Williams, Dionex Corporation, New Jersey, private communication.

212. F. C. Smith, Jr., Microsensor Technology Inc., Fremont, California, unpublished work.

213. F. Karaswa, T. Ono, and S. Mizusawa, "Analysis of Photographic Processing Solutions by Ion Chromatography, I. Analysis of Developing Solutions," *Nippon Shashin Gakkaishi,* **43**(4), 245–250, (1980) (in Japanese).

214. T. Ono, M. Kamoi, F. Karaswa, and S. Mizusawa, "Analysis of (Sodium) Thiosulfate in Photographic Gelatin by Ion Chromatography," *Nippon Shashin Gakkaishi,* **43**(1), 48–50 (1980).

215. "Anions in Photographic Fixer," Wescan Corporation, Santa Clara, California, Ion Analysis Applications No. 7.

216. "Analysis of Engine Coolants Using Ion Chromatography," Dionex Corporation, Sunnyvale, California, Applications Note No. 4.

217. "Routine Determination of Chromate," Application Note No. 26, and "Determination of Sulfate and Chromate in Electroplating Baths," Application Note No. 20, Dionex Corporation, Sunnyvale, California.

218. "Anions in Plating Bath Solution," Wescan Corporation, Santa Clara, California, Ion Analysis Application No. 14.

219. T. R. Dulski, "Determination of Acid Concentrations in Specialty Alloy Pickling Baths by Ion Chromatography," *Anal. Chem.,* **51**(9), 1439–1443 (1979).

220. "Analysis of Oxalate in Industrial Liquors," Dionex Corporation, Sunnyvale, California, Application Note No. 5.

221. "Anions in Aluminum Processing (Bayer) Liquor," Wescan Corporation, Santa Clara, California, Ion Analysis Application No. 12.

222. J. Menear, Dionex Corporation, Ft. Worth, Texas, private communication.

223. R. C. Chang, "Ion Chromatography: A New Technique in Water Analysis," presented at the Pittsburgh Conference on Analytical Chemistry and Applied Spectroscopy, Cleveland, Ohio, March 1977.

224. R. P. Lash and C. J. Hill, "Ion Chromatographic Determination of Dibutylphosphoric Acid in Nuclear Fuel Reprocessing Streams," *J. Liq. Chromatog.,* **2**(3), 417–427 (1979); also J. D. Mulik and E. Sawicki, Eds., *Ion Chromatographic Analysis of Environmental Pollutants,* Volume 2, Ann Arbor Science, Ann Arbor, Michigan, 1979, pp. 67–74.

225. J. M. Keller and R. R. Rickard, "Determination of Dibutylphosphoric Acid in Carbonate, Oxalate, or Nitrate Solutions by Ion Chromatography," *Energy Res. Abstr.,* **6**(6) (1981), Abstract No. 6996.

226. C. J. Hill and R. P. Lash, "Ion Chromatographic Determination of Boron as Tetrafluoroborate," *Anal. Chem.,* **52**(1), 24–27 (1980); see also *Can. Res.,* February 1980.

227. J. D. Sinclair, "Ion Chromatographic Analyses of Contaminants on Zinc and Aluminum Surfaces Exposed to a Range of Urban Environments," presented at the Rocky Mountain Conference, Denver, Colorado, August 1980.

228. W. B. Wargotz, "Ion Chromatographic Quantification of Contaminant Ions in Water Extracts of Printed Wiring," *Surf. Contam: Genesis Detect. Control* (*Proc. Symp.*), Plenum, New York, 1979, pp. 877–895.

229. K. Bonner, "A Comparison of Direct Surface Analysis Technique with Solvent Extraction/Contaminant Profiling Technique for Ascertaining Production Cleanliness in Printed Wiring," *IDC/TP*, 323, April 1980.

230. A. Attia, Gould Corporation, Chicago, Illinois, private communication.

231. L. A. Posta, "Ion Chromatography as a Problem Solving Tool for Electronic Components and Equipment," presented at the Rocky Mountain Conference, Denver, Colorado, August 1980.

232. L. A. Posta, "Further Work in the Use of Ion Chromatography as a Problem-Solving Tool for Electric Components and Equipment," presented at the Rocky Mountain Conference, Symposium on Ion Chromatography, Denver, Colorado, August 1981.

233. J. G. Tarter, K. L. Evans, and C. B. Moore, "A Method for the Ion Chromatographic Determination of Sulfur in Industrial Steels and Iron Meteorites," presented at the Rocky Mountain Conference, Denver, Colorado, August 1980.

234. A. D. Murray, C. J. Vairo, and R. E. Lett, "Application of Ion Chromatography to Geochemical Exploration," presented at the Rocky Mountain Conference, Denver, Colorado, August 1980.

235. D. J. Koop and A. R. Barringer, "Determination of Anions in Geological Samples by Ion Chromatography and Its Application to Uranium Exploration," presented at the Pittsburgh Conference on Analytical Chemistry and Applied Spectroscopy, Cleveland, Ohio, February 1979.

236. K. L. Evans and C. B. Moore, "Combustion–Ion Chromatographic Determination of Chlorine Silicate Rocks," *Anal. Chem.,* **52**(12), 1908–1912 (1980).

237. K. L. Evans, J. G. Tarter, and C. B. Moore, "Pyrohydrolytic Ion Chromatographic Determination of Fluorine, Chlorine, and Sulfur in Geological Samples," *Anal. Chem.,* **53**(6), 925–928 (1981).

238. J. M. Baldwin and P. R. Klock, "Ion Chromatographic Determi-

nation of Non-metallic Elements in Geological Materials," presented at the Rocky Mountain Conference, Symposium on Ion
Chromatography, Denver, Colorado, August 1981.

239. C. B. Moore, K. L. Evans, and T. G. Tarter, "A Rapid Method
for the Ion Chromatographic Determination of Fluorine, Chlorine,
and Sulfur," presented at the Rocky Mountain Conference, Symposium on Ion Chromatography, Denver, Colorado, August 1981.

240. S. H. Peterson, J. C. Bellows, D. F. Pensenstadler, and W. M.
Hickam, "Steam Purity Monitoring for Turbine Corrosion Control," presented at the International Power Conference, Pennsylvania, 1979.

241. W. M. Hickam, J. C. Bellows, S. H. Peterson, D. F. Pensenstadler,
E. K. Evans, C. Green, and E. G. Mallari, "Steam/Condensate
Chemical Maintenance and Control in Power Plant Operations,"
presented at the 41st International Water Conference, Pittsburgh,
Pennsylvania, October 1980.

242. M. A. Fulmer, J. Penkot, and R. Nadlin, "Sodium, Potassium,
Chloride and Sulfate Analysis at ppb Levels in Water," Westing-
house Scientific paper 78-1P4-ACHTA-P2; see also J. D. Mulik and
E. Sawicki, Eds., *Ion Chromatographic Analysis of Environmental
Pollutants*, Vol. 2, Ann Arbor Science, Ann Arbor, Michigan, 1979,
pp. 381–400.

243. S. A. Borman, "Monitoring Steam to Combat Corrosion," *Anal.
Chem.*, **52**(13), 1409–1410 (1980).

244. K. M. Roberts, D. T. Gjerde, and J. S. Fritz, "Single-Column Ion
Chromatography for the Determination of Chloride and Sulfate in
Steam Condensate and Boiler Feed Water," *Anal. Chem.*, **53**(11),
1691–1694 (1980).

245. "Direct Loading Procedure for Trace Analysis in Power Plants,"
Dionex Corporation, Sunnyvale, California, Technical Note 10/80.

246. B. Hickam, J. Bellows, D. Pensenstadler, and S. Peterson, "Determination of Steam Purity in Turbines and Power Plants," Westinghouse Report (1979).

247. J. Brown and R. E. Massey, "Condensate, Feedwater, Steam Sampling and Analysis in Ontario Hydro Thermal Generating Stations,"
presented at the 41st International Water Conference, Pittsburgh,
Pennsylvania, October 1980.

248. G. L. Carlson, "Characterization of the Steam Turbine Chemical
Environment," presented at the Steam Turbine Generator Technology Symposium, Charlotte, North Carolina, October 1978.

249. W. Nestel, W. Lechnick, and R. C. Rice, "Analysis for pH, Chlorides, Dissolved Oxygen, Conductivity and Boron Under 'Post-Accident' Conditions," NUS, presented at Sentry Seminar, February 1980.

250. J. J. Law, "Boiler Water Analysis Using Ion Chromatography," *Power Engineering,* June 1981.

251. "Critical Points to Consider When Using Concentrator Columns," Dionex Corporation, Sunnyvale, California, Technical Note 9.

252. T. Wilhite, EMCON Associates, San Jose, California, private communication.

253. Dionex Corporation, Sunnyvale, California.

254. Dionex Corporation, Sunnyvale, California.

255. J. P. Wilshire, "Analysis of Boron by Ion Chromatographic Exclusion," presented at the Rocky Mountain Conference, Symposium on Ion Chromatography, Denver, Colorado, August 1981.

256. Dionex Brochure on Electrochemical Detection.

257. W. Nestel, W. Lechnick, and R. C. Rice, "Analysis for pH, Chlorides, Dissolved Oxygen, Conductivity, and Boron Under 'Post-Accident' Conditions," NUS, presented at Sentry Seminar, February 1980.

258. S. J. Johnson, "A Versatile Analytical Tool in the Field of Nuclear Fuels Reprocessing and Waste Management," presented at the Pittsburgh Conference on Analytical Chemistry and Applied Spectroscopy, Cleveland, Ohio, 1979.

259. J. M. Keller, "Application of Ion Chromatography to Nuclear Technology Development," presented at the Rocky Mountain Conference, Denver, Colorado, August 1980.

260. R. J. Sironen, "Modification of a Dionex Model 14 Ion Chromatograph to Permit Analyzing Radioactively Contaminated Samples," presented at the Rocky Mountain Conference, Denver, Colorado, August 1981.

261. R. M. Merrill, "The Determination of Boron and Phosphorus in Silicate Glasses Using Ion Chromatography Exclusion," presented at the Rocky Mountain Conference, Symposium on Ion Chromatography, Denver, Colorado, August 1981.

262. L. U. Curfman and S. J. Johnson, "Modification of an Ion Chromatograph for Analysis of Radioactive Samples," Rockwell International, Report No. RHO-SA-110, November 1979.

263. S. M. Kapelner, J. C. Trocciola, and M. S. Freed, "Trace Level Analysis of Anions in High Purity Water," in J. D. Mulik and E.

Sawicki, Eds., *Ion Chromatographic Analysis of Environmental Pollutants,* Vol. 2, Ann Arbor Science, Ann Arbor, Michigan, 1979, pp. 345–360.

264. F. C. Smith, Jr., Microsensor Technology Inc., Fremont, California, unpublished work.

265. W. E. Rich and L. McNeil, Dionex Corporation, Sunnyvale, California, private communication.

266. J. Menear, Dionex Corporation, Ft. Worth, Texas, private communication.

267. "Analysis of Organic and Inorganic Ions in Milk Products," Dionex Corporation, Sunnyvale, California, Application Note 9.

268. "Analysis of Strong and Weak Acids in Coffee Extracts," Dionex Corporation, Sunnyvale, California, Application Note 19.

269. L. D. Davis and A. K. Foltz, "The Determination of Acids in Coffee by Ion Chromatography (IC) and Ion Chromatography with Exclusion (ICE)," presented at the Rocky Mountain Conference, Symposium on Ion Chromatography, Denver, Colorado, August 1981.

270. "Analysis of Organic Acids and Cationic Species in Wine," Dionex Corporation, Sunnyvale, California, Application Note 21.

271. "Phosphate in Cola," Wescan Corporation, Santa Clara, California, Application 3.

272. R. C. Chang, General Electric, Mt. Vernon, Indiana, unpublished work.

273. G. Crook, Dionex Corporation, Denver, Colorado, private communication.

274. J. Linberg, "Determination and Identification of Sulfite in Some Pharmaceutical Compositions Using Ion Chromatography," presented at the Rocky Mountain Conference, Symposium on Ion Chromatography, Denver, Colorado, August 1981.

275. M. Williams, Dionex Corporation, New Jersey, private communication.

276. D. D. Fratz, "Quantitative Determination of Inorganic Salts in Certifiable Color Additives," in E. Sawicki, J. D. Mulik, and E. Wittgenstein, Eds., *Ion Chromatographic Analysis of Environmental Pollutants,* Vol. 1, Ann Arbor Science, Ann Arbor, Michigan, 1978, pp. 169–184.

277. D. D. Fratz, "Ion Chromatographic Analysis of Food, Drug and Cosmetic Color Additives," in J. D. Mulik and E. Sawicki, Eds.,

Ion Chromatographic Analysis of Environmental Pollutants, Vol. 2, Ann Arbor Science, Ann Arbor, Michigan, 1979, pp. 371–380.

278. S. Abrahms, Dynapol, Palo Alto, California, private communication.

279. R. George, Dionex Ltd., Surrey, England, private communication.

280. J. A. Beech, R. Diaz, C. Ordaz, and B. Palmomeque, "Nitrates, Chlorates, and Trihalomethanes in Swimming Pool Water," *Am. Pub. Health Assoc.,* **70**(1), 79–81 (1981).

281. Dionex Brochure on Electrochemical Detection.

282. W. E. Rich, Dionex Corporation, Sunnyvale, California, private communication.

283. F. C. Smith, Jr., Microsensor Technology Inc., Fremont, California, unpublished work.

284. F. C. Smith, Jr., Microsensor Technology Inc., Fremont, California, unpublished work.

285. Dionex Training Course, January 1977.

286. L. W. Little, R. A. Zweidinger, E. C. Monnig, and W. J. Firth, "Treatment Technology for Pesticide Manufacturing Effluents: Atrazine, Maneb, MSMA, Oxyzalin; Appendix E, Analytical Procedure for Determination of MSMA," EPA Report EPA-600/2-80-043, February 1980.

287. R. Posner and A. Schoffman, "The Analysis of Portland Cements by Ion Chromatography," presented at the Pittsburgh Conference on Analytical Chemistry and Applied Spectroscopy, Cleveland, Ohio, March 1980.

288. A. Fitchett, Dionex Corporation, Sunnyvale, California, private communication.

289. D. D. Fratz, "Ion Chromatographic Analysis of Food, Drug, and Cosmetic Color Additives," in J. D. Mulik and E. Sawicki, Eds., *Ion Chromatographic Analysis of Environmental Pollutants,* Vol. 2, Ann Arbor Science, Ann Arbor, Michigan, 1979, pp. 371–380.

290. V. Smith, Dionex Corporation, Sunnyvale, California, private communication.

291. S. W. Babulak, "The Determination of Sodium Monofluorophosphate and Sodium Fluoride in Dental Cream Using Ion Chromatography," presented at the Rocky Mountain Conference, Denver, Colorado, August 1980.

292. A. Fitchett, Dionex Corporation, Sunnyvale, California, private communication.

293. J. W. Whittaker and P. R. Lemke, "Application of Ion Chromatography to Determination of Inorganic Ions in Pharmaceuticals," presented at the Pittsburgh Conference on Analytical Chemistry and Applied Spectroscopy, March 1980.

294. J. F. Colaruotolo and R. S. Eddy, "Determination of Chlorine, Bromine, Phosphorous, and Sulfur in Organic Molecules Using Ion Chromatography," *Anal. Chem.*, **49**(6), 884–885 (1977).

295. F. Smith, Jr., A. McMurtrie, and H. Galbraith, "Ion Chromatographic Determination of Sulfur and Chlorine Using Milligram and Submilligram Sample Weights," *Microchem. J.*, **22**(1), 45–49 (1977).

296. J. F. Colaruotolo, "Organic Elemental Microanalysis by Ion Chromatography," in E. Sawicki, J. D. Mulik, and E. Wittgenstein, Eds., *Ion Chromatographic Analysis of Environmental Pollutants*, Vol. 1, Ann Arbor Science, Ann Arbor, Michigan, 1978, pp. 149–168.

297. F. Smith, Jr., A. McMurtrie, and H. Galbraith, "Ion Chromatographic Determination of Sulfur and Chlorine Using Milligram and Submilligram Sample Weights," *Microchem. J.*, **22**(1), 45–49 (1979).

298. F. Scheidl, Hoffman La-Roche, Nutley, New Jersey, private communication.

299. F. Smith, Jr., A. McMurtrie, and H. Galbraith, "Ion Chromatographic Determination of Sulfur and Chlorine Using Milligram and Submilligram Sample Weights," *Microchem. J.*, **22**(1), 45–49 (1977).

300. G. Thomas, Institute of Mining and Metallurgy Research, Lexington, Kentucky, private communication.

301. D. V. Vinjamoori and C. Ling, "Personal Monitoring Method for Nitrogen Dioxide and Sulfur Dioxide with Solid Sorbent Sampling and Ion Chromatographic Determinations," *Anal. Chem.*, **53**(11), 1689–1690 (1981).

302. L. C. Speitel, J. C. Spurgeon, and R. A. Filipczak, "Ion Chromatographic Analysis of Thermal Decomposition Products of Aircraft Interior Materials," in J. D. Mulik and E. Sawicki, Eds., *Ion Chromatographic Analysis Environmental Pollutants*, Vol. 2, Ann Arbor Science, Ann Arbor, Michigan, 1979, pp. 75–88.

303. R. D. Holm and S. A. Barksdale, "Analysis of Anions in Combustion Products," in E. Sawicki, J. D. Mulik, and E. Wittgenstein, Eds., *Ion Chromatographic Analysis of Environmental Pollutants*,

Vol. 1, Ann Arbor Science, Ann Arbor, Michigan, 1978, pp. 99–110.

304. F. E. Butler, F. J. Toth, D. J. Driscoll, J. N. Hein, and R. H. Jungers, "Analysis of Fuels by Ion Chromatography: Comparison with ASTM Methods," in J. D. Mulik and E. Sawicki, Eds., *Ion Chromatographic Analysis of Environmental Pollutants*, Vol. 2, Ann Arbor Science, Ann Arbor, Michigan, 1979, pp. 185–192.

305. R. Windham, Jefferson Chemical, Austin, Texas, private communication.

306. C. Anderson, "Ion Chromatography: A New Technique for Clinical Chemistry," *Clin. Chem.*, **22**(9), 1424 (1976).

307. C. Mahle and M. Menon, "Oxalate Quantitation by Ion Chromatography: Clinical Applications," presented at the Rocky Mountain Conference, Symposium on Ion Chromatography, Denver, Colorado, August 1981.

308. W. Rich, E. Johnson, L. Lois, P. Kabra, B. Stafford, and L. Marton, "Determination of Organic Acids in Biological Fluids by Ion Chromatography: Plasma Lactate and Pyruvate and Urinary Vanillylmandelic Acid," *Clin. Chem.* (Winston-Salem, North Carolina), **26**(10), 1492–1498 (1980).

309. W. E. Rich, E. Johnson, L. Lois, B. E. Stafford, P. M. Kabra, and L. J. Marton, "Clinical Analysis of Endogenous Human Biochemicals: Organic Acids by Ion Chromatography," *Liq. Chromatogr. Clin. Anal.*, 393–407, (1981).

310. S. J. Rehfeld, H. F. Laken, F. R. Nordmeyer, and J. D. Lamb, "Improved Ion-Chromatographic Method for Determining Magnesium Ion and Calcium Ion in Serum," *Clin. Chem.* (Winston-Salem, North Carolina), **26**(8), 1232–1233 (1980).

311. A. MacDonald, R. Saperstein, F. Scheidl, E. Hargroves, and A. Love, "Ion Chromatographic Analysis of Animal Feeds," presented at the Rocky Mountain Conference, Symposium on Ion Chromatography, Denver, Colorado, August 1981.

312. D. E. C. Cole and C. R. Scrivner, "Sulfate Assay in Biological Fluids by Ion Chromatography," presented at the Rocky Mountain Conference, Symposium on Ion Chromatography, Denver, Colorado, August 1981.

313. W. Rich, F. Smith, Jr., L. McNeil, and T. Sidebottom, "Ion Exclusion Coupled to Ion Chromatography: Instrumentation and Application," in J. D. Mulik and E. Sawicki, Eds., *Ion Chromatographic Analysis of Environmental Pollutants*, Vol. 2, Ann Arbor Science, Ann Arbor, Michigan, 1979, pp. 17–30.

314. General Brochure, The Separations Group.

315. L. C. Westwood and E. L. Stokes, "Determiniation of Azide in Environmental Samples by Ion Chromatography," in J. D. Mulik and E. Sawicki, Eds., *Ion Chromatographic Analysis of Environmental Pollutants,* Vol. 2, Ann Arbor Science, Ann Arbor, Michigan, 1979, pp. 141–156.

316. R. K. Pinschmidt, "Ion Chromatographic Analysis of Weak Acid Ions Using Resistivity Detection," in J. D. Mulik and E. Sawicki, Eds., *Ion Chromatographic Analysis of Environmental Pollutants,* Vol. 2, Ann Arbor Science, Ann Arbor, Michigan, 1979, pp. 41–50.

317. R. K. Pinschmidt and M. J. Fasolka, "Resistivity Detection of Weak Acids: Carbamates and Carbonates," presented at the Rocky Mountain Conference, Symposium on Ion Chromatography, Denver, Colorado, August 1981.

318. J. E. Girard, "Ion Chromatography with Coulometric Detection for Determination of Inorganic Ions," *Anal. Chem.,* 51(7), 836–839, (1979).

319. J. E. Girard, "Ion Chromatography with Coulometric Detection for Determination of Inorganic Ions," *Anal. Chem.* 51(7), 836–839, (1979).

320. Dionex Brochure on Electrochemical Detection.

321. L. C. Westwood, Ford Motor Company, Dearborn, Michigan, private communication.

322. R. Williams, "Application of Ion Chromatography to Analysis of Organic Sulfur Compounds," presented at the Rocky Mountain Conference, Denver, Colorado, August 1980.

323. D. A. Otterson, "Application of Ion Chromatography to the Study of Hydrolysis of Some Halogenated Hydrocarbons at Ambient Temperatures," *NASA Tech. Memo,* NASA-TM-P1020, E-9817 (1978), 24 p.

324. F. R. Nordmeyer, C. D. Hansen, D. J. Eatough, D. K. Rollins, and J. D. Lamb, "Determination of Alkaline Earth and Divalent Transition Metal Cations by Ion Chromatography with Sulfate-Suppressed Barium and Lead Eluents," *Anal. Chem.,* 52(6), 852–856 (1980).

325. J. D. Lamb, L. D. Hansen, G. G. Patch, and F. R. Nordmeyer, "Iodate-Suppressed Lead Eluent for Ion Chromatographic Determination of Divalent Cations," *Anal. Chem.,* 53(4), 749–750 (1918).

326. E. L. Johnson and E. Cathers, "Applications of Simultaneous Detection in IC," presented at the Rocky Mountain Conference, Symposium on Ion Chromatography, Denver, Colorado, August 1981.

327. N. S. Simon, "The Removal of Iron Interferences in IC Analyses by Complexation," presented at the Rocky Mountain Conference, Symposium on Ion Chromatography, Denver, Colorado, August 1981.

328. C. A. Pohle and M. Ebenhahn, "Mobile Phase Ion Chromatography (MPIC)," presented at the Rocky Mountain Conference, Symposium on Ion Chromatography, Denver, Colorado, August 1981.

329. J. Slanina and F. P. Bakker, "Development of High Speed Ion Chromatography," presented at the Rocky Mountain Conference, Symposium on Ion Chromatography, Denver, Colorado, August 1981.

330. C. A. Pohl and A. Woodruff, "High Performance Ion Chromatography (HPIC)," presented at the Rocky Mountain Conference, Symposium on Ion Chromatography, Denver, Colorado, August 1981.

331. J. Slanina and F. P. Bakker, "Development of High Speed Ion Chromatography," presented at the Rocky Mountain Conference, Symposium on Ion Chromatography, Denver, Colorado, August 1981.

CHAPTER
7
Methods Development in IC

Standard IC operating conditions are expanded as work is done to refine these analytical techniques. Quite a number of alternatives to the original sets of standard operating conditions have been developed and are covered in the chapters on IC practice and applications. Standard operating conditions should always be tried first when a new or difficult separation is attempted. This will give the analyst some important clues to guide methods development. New ion chromatographic separation and detection methods must be developed for two reasons. First, if the conventional operating modes are inadequate to perform the required ion analyses, new methods must be sought. Second, new procedures are sometimes needed for ions or complex sample matrices that had not been analyzed previously. In any event, whether new ions are being determined or a new type of sample is being tried for the first time, *both* trial and error experimentation *and* some thinking must be done if a new method is to be developed successfully. Once again the best starting point is standard instrument conditions. If they are unsuccessful, the instrument conditions are changed to a different configuration and the sample is reinjected onto this set of IC columns or an entirely different analytical scheme is thought out before the sample is injected. Both steps must be completed to develop a new analytical method. The sequence is left to the analyst.

The approach to methods development presented here has been successful for the authors. An actual case in which a method of analysis for caustic samples was developed for IC by questioning procedures and IC experiments is used as an example.[1] The methods development sequence starts with several lists of questions about the sample and how it relates to ion chromatographic analysis. Again, the first step in any analysis is to try standard IC conditions. The example is carried through the various stages of IC analysis that lead to answers for the four separations problems. The results of these analyses are then detailed. This procedure actually lead to the development of coupled chromatography.

7.1   THE INITIAL ANALYTICAL PROBLEM

The first set of questions about the sample has little to do with ion chromatography. The overall analytical problem must be carefully defined before work can begin on the samples. Among the questions that need to be answered by the analyst are the following:

1. Which ions need to be determined?
2. What analysis methods are available for these determinations?
3. What are the approximate concentrations of these ions?
4. What other species are or may be present?
5. What are the levels of these other species?
6. How many samples are there?
7. In what time frame must these analyses be done?
8. What accuracy and precision levels are acceptable?
9. How do these ions behave in strong acids and strong bases (used as eluents in IC) or in the suppressor resin?
10. Will these ions interact with the resin polymer?
11. What nonionic species are present?
12. Are special handling procedures or instrumental methods necessary?

The answers to these questions indicate whether ion chromatography is potentially useful in the analysis of a particular sample set.

For the caustic sample the ions of interest are carbonate, chloride, hypochlorite, chlorate, sulfate, calcium, and magnesium. The matrix is 50% NaOH; Na^+ and OH^- are the main ionic constituents. Trace levels of transition metals and organic acids may also be present in the sample. The hydroxide will neutralize part of the H^+ cation eluent and act as an eluting ion in the anion determinations. The samples should not be exposed to air; they should be stored in plastic containers and carefully handled by all laboratory personnel.

7.2 INITIAL IC TESTING

Now that the analytical requirements have been established, they must be compared with the capabilities and limitations of the ion chromatograph. Once again a series of questions can provide a reasonable evaluation of the possible uses of IC. Questions that need to be answered are the following:

1. Have these ions ever been analyzed by IC?
2. Have similar species been separated by IC?
3. What is the pK_A or pK_B of the ions of interest?
4. Are there high concentrations of ions or major ionic components that will result in column overloading?
5. If the pK of the ions of interest is greater than 5, can these species be converted to more conductive forms?
6. Can chemical pretreatments be applied to minimize potential interference or overloading?
7. Are the ions electroactive or do they absorb UV or visible light?

The methodology listing in Chapter 3 should be used with published papers on IC and ion exchange (especially affinity tables or retention tables) to guide the analyst in the selection of IC instrument conditions for the initial analysis. When the method was developed for analysis of caustics the chloride, chlorate, sulfate, calcium, and magnesium ions had already been determined by IC but not in high NaOH solutions. Coupled chromatography had not yet been developed. The hypochlorite and carbonate ions had not been successfully separated or detected by IC. The high OH^- concentration is expected to cause problems with anion separations and high Na^+ concentrations will make cation analyses difficult. The carbonate ion may be difficult to detect at low levels. Neutralization of the hydroxide (OH^-) with acid will add a large amount of another anion (Cl^-, NO_3^-, or SO_4^{2-}) that will have an unpredictable effect on the resulting ion chromatogram.

At this point two options exist for the analyst. The sample can be injected into the IC and the chromatogram analyzed, or a possible chromatogram can be constructed from a third set of questions directly related to IC analyses. These questions call on the knowledge and experience of the IC operator:

1. Is the sample aqueous or can it be made aqueous?
2. Are analyses for cation, anions, or both required by IC?
3. Is sample pretreatment by dissolution or chemical/physical means necessary?
4. Should the sample be diluted? By how much and with eluent or water?
5. Are there suspended solids that need to be filtered?

6. What is the range of relative affinities of the sample ions for the resin?

7. Which column sets should be used?

8. Which eluents should be used? Is isocratic elution acceptable?

For the 50% NaOH example the sample must be analyzed for anions and divalent cations. The starting point for each of these analyses should be standard operating conditions. The high NaOH concentration makes at least a 10-fold dilution necessary. The other analytical requirements (e.g., the need to determine CO_3^{2-}) make water the only acceptable diluent. Figures 7.1 and 7.2 are chromatograms for the initial injections of the 10-fold diluted sample obtained under cation and anion standard conditions. The sodium level in Figure 7.1 results in a peak that is too broad to allow sensitive detection of Ca^{2+} and Mg^{2+}. The anion chromatogram shows that the peaks tentatively identified as ClO_3^- and SO_4^{2-} elute faster than these ions do in a synthetic standard (Figure 7.3). No separation of hypochlorite and chloride are observed and a broad peak eluting at about the same time as chloride is noted. Individual spiking with ClO^-, Cl^-, and CO_3^{2-} standards verified that the broad peak is probably carbonate

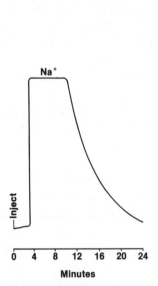

Figure 7.1.   Cations in 50% NaOH.

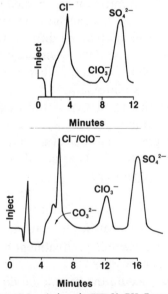

Figure 7.2.   Anions in 50% NaOH. Copyright by Dionex Corp. and reprinted by permission of copyright owner.

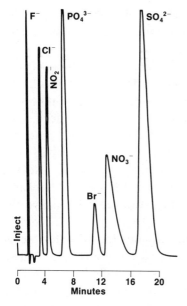

Figure 7.3. Typical inorganic anion separation. Copyright by Dionex Corp. and reprinted by permission of copyright owner.

and that hypochlorite and chloride are not likely to be resolved by standard conditions. Thus it appears that standard IC conditions are inadequate to analyze these samples.

7.3   AFTER INITIAL IC SEPARATIONS FAIL TO GIVE ADEQUATE RESULTS

From the preceding two chromatograms four separate methods development problems exist:

1. Separation of magnesium and calcium in high sodium concentrations.
2. Separation and quantitation of carbonate ion.
3. Solution of the peak shift problem for chlorate and sulfate.
4. Separation of chloride and hypochlorite.

Each of these problems must be attacked individually. All knowledge of ion chemistry, ion exchange, or chromatography must be applied to obtain solutions. The chapter on IC practice in this book provides a good starting point (keep in mind that coupled chromatography had not been developed). Books or papers on ion exchange should also be sought out. Inorganic chemistry texts are valuable when planning the alternate courses of action required in a methods development exercise.

The separation of magnesium and calcium ions when high concentrations of sodium ion are present is quite difficult. Larger separator columns will separate these divalent ions. Experiments with a 9 × 250 mm cation separator and two 6 × 250 mm cation separator columns produced some separation but the Ca^{2+} and Mg^{2+} peaks eluted quite late. The broadness of these peaks makes trace detection of these species impossible, even at the upper concentration limit expected for Ca^{2+} in the 50% NaOH. Further dilution of the sample should result in poorly detectable peaks; therefore it was not tried. Another approach was needed to solve the problem.

Because Ca^{2+}/Mg^{2+} have much higher affinities for the cation separator resin than Na^+, a method that involves concentration of the divalent cations while minimizing Na^+ concentration seemed the best approach. One milliliter of the 10-fold diluted sample was passed slowly through a 3 × 150 mm cation precolumn. It was hoped that the Ca^{2+}/Mg^{2+} would be retained by the resin and excess Na^+ ions would remain in the void volume of the column. This sample was then given a 3-mL wash with DI water to remove excess Na^+ ions. Figure 7.4 is the resulting chromatogram. The Na^+ concentration was low enough to allow normal separation of Ca^{2+} and Mg^{2+}. A water blank shows small Ca^{2+}/Mg^{2+} peaks.[2] Calcium and magnesium recoveries were not studied.

The second problem was the elution and sensitive detection of carbonate ion. The broad CO_3^{2-} peak in the standard anion chromatogram increased in width with suppressor column length, thus indicating a possible non-ion exchange interaction in the suppressor column. Use of a longer anion separator column was not effective in obtaining better Cl^-/CO_3^{2-} separation. Neither was the use of a smaller suppressor column. Because the carbonate ion apparently interacted with the suppressor resin, experiments were begun to determine whether Cl^- and CO_3^{2-} could be separated with a suppressor-type resin. A lower cross-linked cation exchange resin was packed into a 6 × 500 mm glass column. Water and $Na_2B_4O_7$ were tried as eluents; both seemed to work. No suppressor column was used. Figure 7.5 shows the elution of carbonate ion in 50% NaOH by direct injection. The neutralization of the OH^- by the resin caused the column to become warm and to discolor temporarily. The CO_3^{2-} peak is quite reproducible. The carbonate ion was verified by spik-

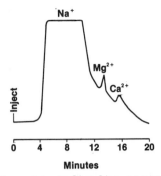

Figure 7.4. Mg²⁺/Ca²⁺ in high Na⁺ solution after concentration step. Copyright by Dionex Corp. and reprinted by permission of copyright owner.

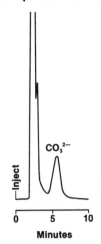

Figure 7.5. Carbonate ion in 50% NaOH. Copyright by Dionex Corp. and reprinted by permission of copyright owner.

ing and conventional titration of the trapped peak. The chloride, hypochlorite, chlorate, and sulfate all eluted in the void volume peak of the ion exclusion column.

The third problem was much easier to solve than the first two. Dilution of the sample with anion eluent virtually eliminated the peak shifts for ClO_3^- and SO_4^{2-}. Determination of CO_3^{2-}, however, now required another injection or a dilution with preboiled water. Another pretreatment to remove the OH^- ions was preferred. Addition of a strong acid to neutralize the hydroxide introduced a large concentration of another anion, which complicates the anion separation even further. A better method was devised where the sample is passed through an ion exchange column in the H^+ form. The ClO_3^- and SO_4^{2-} ions were eluted together in the void volume of this column; they were then trapped on an anion concentrating column for subsequent analysis by standard anion ion chromatography.

The solutions to the problems of carbonate/chlorate/sulfate separations led to the coupled chromatography concept. In this method the 10-fold diluted sample is injected through an ion exclusion column to separate the strong acid species from the weak acids and to neutralize the OH^- ion. A concentrating column is used to trap the void volume peak which contains ClO^-, Cl^-, ClO_3^-, and SO_4^{2-}. These ions are then analyzed by standard anion IC. Figure 7.6 shows a typical coupled separation for a weak (4.8%) caustic solution. All the quirks have not been eliminated from this system, however; when 50% NaOH is directly injected into the

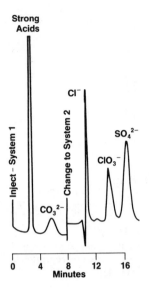

Figure 7.6. Separation of anions in 4% caustic. Copyright by Dionex Corp. and reprinted by permission of copyright owner.

IE system the carbonate peak is constant in peak height, but the chloride/hypochlorite, chlorate, and sulfate peak heights increase with each subsequent injection. Aqueous standards show excellent reproducibility when run after a strong caustic injection. This problem is not so severe when 5% NaOH solutions are used. Further work needs to be done on this problem.

The fourth problem, the separation of ClO^- and Cl^-, has just recently been solved by ion chromatography. Longer separating columns, different eluents, different separating resins, and several other methods had not led to acceptable separations. Sample prechemistry with arsenite ion and thiosulfate had shown some promise, but neither solution seems to have been successful in obtaining reasonable chloride values. Use of silver suppression of the chloride ion would result in the formation of explosive $AgClO$; therefore this method has not been pursued. The best method available utilizes an electrochemical detector with a glassy carbon electrode to detect the ClO^- ion selectively after ion exchange separation. This method had just recently been shown when this section was written. Results for Cl^-/ClO^- separation appear to be simple to obtain when electrochemical and conductivity detection are used concurrently.

From this extensive, long-term development of an optimized caustic analysis the best approach at the time of this writing was to separate carbonate and other weak acids, as well as chlorate and sulfate, by coupled chromatography. Magnesium and calcium ions require a special procedure that concentrates these ions as it removes excess sodium. This method has not yet been finalized and recovery studies are not yet thorough enough to validate this procedure for routine analytical use. In both cases a 10-fold dilution with carbonate-free water is preferred. The ion chromatographic analysis of a caustic sample now requires about ½ h, whereas wet chemical methods required 2 to 4 h per sample.

7.4 OTHER METHODS OF SOLVING ION ANALYSIS PROBLEMS

Several aspects must be carefully considered when the development of a new ion chromatographic-based method is planned. Suppressor-only injections and standard operating conditions with diluted samples (100 to 1000 times) are usually the best starting points for an initial analysis. After the initial run species that are not detectable when injected through the suppressor column will require different detection systems. Species that elute rapidly with standard conditions will require longer columns, weaker strength eluents, or weaker eluting ions. Ions that are strongly retained on the separator column must be run with shorter separators, stronger eluting ions, or higher strength eluents. Other separating systems and detectors should always be considered when the solution of a difficult ion analysis problem is attempted.

Other methods of solving specific separations or detection problems have been developed around ion chromatographic resins and detectors. Resistivity (negative peak) detection has been applied to the analysis of weakly ionized species,[3] and electrochemical detection has also been used for these ions.[4,5] Boron has been analyzed as BF_4^{2-} by conventional ion chromatography after the boron species were converted to BF_4^{2-} on an anion exchange resin.[6] Transition metals have been detected by barium eluents and sulfate suppression[7] or by complexing the metals as tartrates and then using an electrochemical detector.[8] Polyphosphates have been separated by sodium citrate eluents.[9] Post column reaction with para-aminoresorcinol and UV-visible detection is also useful for determination of metal ions. Other novel approaches to extend the ion chromatographic technique will be developed in the future.

NOTES

1. Original question concept courtesy of A. Fitchett, Dionex Corporation, Sunnyvale, California, private communication.

2. F. C. Smith, Jr., V. T. Smith, and J. K. Ung, "Optimized Analysis of Brine and Caustic Samples by Ion Chromatography," presented at Pittsburgh Conference on Analytical Chemistry and Applied Spectroscopy, Atlantic City, New Jersey, March 1980.

3. R. K. Pinschmidt, Jr., "Ion Chromatographic Analysis of Weak Acids Using Resistivity Detection," in J. D. Mulik and E. Sawicki, Eds., *Ion Chromatographic Analysis of Environmental Pollutants,* Vol. 2, Ann Arbor Science, Michigan, 1979, pp. 41–50.

4. J. E. Girard, "Ion Chromatography with Coulometric Detection for Determination of Inorganic Ions," *Anal. Chem.,* **51**(7), 836–839 (1979).

5. W. Rich, E. Johnson, L. Lois, P. Kabra, B. Stafford, and L. Marton, "Determination of Organic Acids in Biological Fluids by Ion Chromatography: Plasma Lactate and Pyruvate and Urinary Vanillylmandelic Acid," *Clin. Chem.,* **26**, 1492–1498 (1980).

6. C. J. Hill and R. P. Lash, "Ion Chromatographic Determination of Boron as Tetrafluoroborate," *Anal. Chem.,* **52**(1), 24–27 (1980).

7. L. D. Hansen, J. F. Ryder, N. F. Mangelson, N. W. Hill, K. J. Faucet, and D. J. Eatough, *Anal. Chem.,* **52**, 821 (1980).

8. J. E. Girard, "Ion Chromatography with Coulometric Detection for Determination of Inorganic Ions," *Anal. Chem.,* **51**(7), 836–839 (1979).

9. F. C. Smith, Jr., Microsensor Technology Inc., Fremont, California, unpublished work.

CHAPTER
8
Future Ion Chromatography Developments

As an analytical method, ion chromatography is now emerging from its early stages of development. Existing instrumentation and IC methods in many ways resemble current liquid chromatographic techniques and equipment. Because IC is a specialized subset of HPLC methods, we expect the major evolution of IC to follow the path of other HPLC techniques.

In the last few years HPLC instrumentation has dramatically matured. Recent developments in liquid chromatographic instruments have centered in three main areas: (1) more sophisticated pumping systems, (2) improved column and eluent technology, and (3) improved detectors. Great improvements have been made in the HPLC pump ability to deliver "constant," nearly pulseless eluent flow against operating pressure drops as high as 10,000 psig. Microprocessor-controlled dual or triple piston pumps capable of a wide variety of step or continuous gradients are now available. This innovation has allowed factors such as solvent compressibility to be factored into gradients. Ternary solvent delivery systems have enhanced the gradient capabilities of HPLC equipment. Along with vastly improved pumps, separation technology has evolved rapidly. Use of small-particle size reverse-phase silica packings has enhanced the separation efficiency of HPLC columns. These smaller diameter particles have forced the pump improvements. Narrow particle distribution, advanced column and end-fitting design, and improved consistency in the manufacturing of the packings have resulted in more reproducible separations. These smaller particle packings often have high theoretical plate counts. With the improvements in column packing, solvents have also been improved to optimize separations. Reverse-phase methods with high purity solvents in a variety of isocratic and gradient modes have resulted in a vast expansion in species that can be separated by HPLC methods. Detectors have also evolved significantly. They are more versatile, more reliable, have better sensitivity, and are less expensive. The variety of detectors available has also expanded. Refractive index, ultraviolet (both fixed and variable wavelength), electrochemical, and other detection systems have been used for sensitive, in-line detection of separated species.

Because ion chromatography is a specialized subset of liquid chromatography, it should also develop along lines parallel to HPLC. The use of Tefzel tubing instead of Teflon tubing allows the IC to be used as a specialized HPLC capable of operating in the 1500 to 2000 psig range like other HPLC equipment. Other injection and switching valves must be used when system pressures reach these levels. Higher efficiency columns will result from improved column design and from smaller sized resin particles used as packings. This will mandate higher system-operating pressures. Probably more significant are the resins that have been developed to change ion-exchange selectivities. This results in many improved or some totally new separation capabilities. Indeed, the SA-2 columns sold by Dionex appear to have a different the amine functional group on the anion separator resin latex. Other column systems with different separating columns, different ion exchange functional groups, or smaller particle sizes have been introduced. Improved detectors, coupled detectors, or flow injection will be used to enhance the specificity or increase the analytical range of IC detector systems. These detectors will include electrochemical, photometric, refractive index, improved conductivity, spectroscopic and in-line ion selective electrode detectors as well as inductively coupled plasma, atomic absorption, and other novel detection systems. Many of these detectors have already been used with ion chromatography separations. Single-column, nonsuppressed methods should improve substantially as different eluents and improved separating resins make their appearance.

Several instruments quite different from general laboratory analyzers are expected to emerge in the near future. The expected increase in the use of coupled chromatography for on-line sample treatment and specificity of ion separations will lead to more versatile instrumentation. These systems will have components designed to span a variety of analytical needs. Future ion chromatographs will range from portable, low-cost units for field use to sophisticated research instruments that incorporate gradient elution, considerable microprocessor-based automation, temperature control, and several column systems and detectors in a single unit. Process instrumentation for on-line monitoring in plant locations will feature columns and eluents specifically tailored to one analytical need.

In addition to instrument changes, major developments in columns and eluents are expected. Total analysis time per sample has been reduced significantly without loss of resolution. This will make the ion chromatograph more competitive with traditional "number crunchers" such as automated wet chemical and ion selective electrode methods for the analysis of single ions. Investigation of new eluting species, increased use of organic eluent modifiers and organic solvents in the eluents and temperature control should lead to some interesting separation capabilities in existing and new resins. These improvements will lead IC to the analysis of more complex ions than is presently done. As the various separation mechanisms that take place in styrene–divinylbenzene resins and silica-based systems are carefully studied and better understood, new solvent systems, eluent additives, and separating systems should emerge. These separating systems will be resins with improved selectivity from pairing knowledge of classical ion-exchange properties with liquid chromatography principles. Gradient elution will probably be done with step gradients as long as conductivity is the main detector used in IC. Continuous gradients will be used with other detectors that do not respond to elucent changes. Ion chromatographic separations based on inorganic ion exchangers, liquid/liquid exchangers, paired ion chromatography and other mechanisms will be developed to solve problems in ion analysis.

Clearly, the future of ion analysis by chromatographic methods offers great potential. The range of ions determinable by IC is expanding rapidly to include weakly ionized species. Rapid progress in expanding the field of ion chromatgraphy should make possible more rapid, more specific ion analyses by combining better hardware, improved detectors, more efficient and selective columns, and new eluents and solvents from gradient elution schemes.

The scope of ion chromatography has changed dramatically in the last year and is presently expanding rapidly. The present meaning of IC encompasses ion-exchange resin separations coupled to conductivity detection and electrochemical/UV detectors. Detection of simple ions is emphasized. This will soon expand to broader, more accurate meaning for the term ion chromatography: that is to say, "the chromatographic determination of ionic species with in-line detection."

Appendix

$\mathbf{T}$he ion chromatograph is used to determine a variety of anions and cations, both organic and inorganic. One of the major features of IC is that several ions can be determined in a single injection. This feature also leads to the problem of peak interferences. Two main types of interference are observed in IC: coeluting or nearly coeluting peaks and column overloading from one major ion. This appendix is intended to summarize the capabilities of IC separations which are based on the affinity of the sample ions and eluent ions for the active sites of the resin.

Numerous sources of information have been used to compile the elution tables in this appendix. Literature from instrument manufacturers, scientific publications, and our own unpublished works are the primary sources.

Most of the data in these tables refer to standard IC operating conditions unless otherwise specified. When different conditions are used to produce better separations the eluent and separating column are given. The ions elutable by IC have been arbitrarily broken up by several divisions. First, the ions are classified as anions or cations. Each group is then broken down according to the relative affinities of the various ions for the separating resin. New ions, especially organic ions, are continually being added to these listings. Also, a short list of ions which have not yet been separated by IC or ICE is included. This list is becoming shorter.

Table A.1. IC Index of Inorganic Halogen Anions

Anion	Symbol or Formula	Relative Affinity
Fluoride	F^-	Low
Chloride	Cl^-	Low
Hypochlorite	ClO^-	Low
Chlorite	ClO_2^-	Low
Chlorate	ClO_3^-	Moderate
Perchlorate	ClO_4^-	Very high
Bromide	Br^-	Moderate
Bromate	BrO_3^-	Low
Iodide	I^-	High
Iodate	IO_3^-	Low

Table A.2. IC Index of Inorganic Sulfur Anions

Anion	Symbol or Formula	Relative Affinity
Sulfide	S^{2-}	Low
Sulfite	SO_3^{2-}	Moderate
Sulfate	SO_4^{2-}	Moderate
Thiosulfate	$S_2O_3^{2-}$	High
Thiocyanate	SCN^-	High
Dithionate	$S_2O_6^{2-}$	Very high

Table A.3. IC Index of Inorganic Nitrogen and Boron Anions

Anion	Symbol or Formula	Relative Affinity
Cyanide	CN^-	Low
Azide	N_3^-	Moderate
Nitrite	NO_2^-	Moderate
Nitrate	NO_3^-	Moderate
Borate	BO_3^-	Low
Tetraborate	$B_4O_7^{2-}$	Low
Tetrafluoroborate	BF_4^{2-}	High

Table A.4. Index for Other Inorganic Anions

Anion	Symbol or Formula	Relative Affinity
Arsenate	AsO_4^{3-}	High
Chromate	CrO_4^{2-}	High
Rhenate	ReO_4^{2-}	High
Tungstate	WO_4^{2-}	High
Arsenite	AsO_3^{2-}	Moderate
Phosphonates	—	Very high
Molybdate	MoO_4^{2-}	High
Phosphite	PO_3^{3-}	Moderate
Phosphate	PO_4^{3-}	Moderate

Table A.5. IC Index for Inorganic Cations

Cation	Relative IC Affinity
Lithium	Low
Sodium	Low
Potassium	Moderate
Rubidium	Moderate
Cesium	Moderate
Magnesium	High
Calcium	High
Strontium	High
Barium	Very high
Transition metals (Cu^{2+}, Fe^{3+} etc.)	High

Table A.6. IC Index for Nitrogen Organic Cations

Cation	Relative IC Affinity
Ammonium	Low
Ethanolamine	Low
Isopropanolamine	Low
Triethanolamine	Low-moderate
Methylamine	Moderate
Dimethylamine	Moderate
Trimethylamine	Moderate
n-Butylamine	High
Tetramethylammonium bromide	Moderate
Tetraethylammonium bromide	High
Cyclohexylamine	High-very high
Morpholine	Moderate
Ethylenediamine	Very high
Toluenediamine	Very high
Hydrazine	Low
Guanadine	Low
Nitroguanadine	Low-moderate

Table A.7. IC and ICE Index of Organic Acids

Anion	Relative IC Affinity	Relative ICE Affinity
Acetate	Low	High
Adipate	Moderate	Low-moderate
Acrylate	—	Moderate
Ascorbate	High	—
Benzoate	Moderate	—
Butyrate	Moderate	High
Butyl phosphate	Moderate	—
Citrate	High	Moderate
Chloroacetate	Low	Moderate
Chloropropyl sulfonate	Low-moderate	—
Dibutyl phosphate	Low	—
Dichloroacetate	Moderate	Moderate-high
Ethylmethyl phosphonate	Low	Moderate
Formate	Low	High
Fumerate	Moderate	High
Gluconate	—	High
Glutarate	Moderate	High
Itaconate	High	—
Lactate	Low	High
Levulinate	Low	High
Malate	—	Moderate
Maleate	Moderate	Moderate
Malonate	Moderate	Moderate
Methacrylate	Low	—
Methyl phosphonate	Low	Moderate-high
Methyl isopropyl phosphonate	Low	High
Oxalate	High	Low
Phthalate	High	—
Propionate	Moderate	High
Pyruvate	Low	Moderate
Succinate	Moderate	High
Tartrate	Moderate	High
Trichloroacetate	High	—
Quinate	—	Moderate
Hydroxy-isovalerate	—	High
Hydroxy-*n*-valerate	—	High
Keto-isovalerate	—	Moderate
Cis-oxalacetate	—	Moderate
Ketogluarate	—	Moderate
Acetoacetate	—	High
Gycolate	Low	—

Table A.8. Elution of Weakly Held Anions with Standard Anion IC and Anion ICE Operating Conditions

Anion	IC Conditions				ICE Conditions	
	Approximate Elution Time (min)	Relative Retention[a] (A)	Estimated MDL (ppm)	Other Reported Useful IC Eluents	Approximate Elution Time (min)	Relative Retention (B)
Fluoride	2.5	0.05	0.02	None	9.5	—
Chloride	4.0	0.67	0.05	0.0015M NaHCO$_3$	Precipitates	
Hypochlorite	4.0	0.67	(0.1)?	None	Forms explosive salt	
Chlorite	2.5	0.05	(0.2)?	None	?	?
Bromate	3.5	0.46	(0.2)?	0.0015M NaHCO$_3$	?	?
Iodate	2.5	0.05	(0.3)?	None	?	?
Acetate	2.8	0.17	0.1	0.0025M Na$_2$B$_4$O$_7$	23.8	1.5
Formate	3.2	0.33	0.1	0.0025M Na$_2$B$_4$O$_7$	21.7	1.3
Chloroacetate	3.5	0.46	0.2	0.0015M NaHCO$_3$	NR	NR
Dibutylphosphate	3.0	0.25	0.1	0.003M Na$_2$CO$_3$ with 0.001M NaOH; 0.003M NaOH	NR	NR

Glycolate	3.0	0.25	0.2	0.0015M NaHCO$_3$	NR	NR
Lactate	3.5	0.46	0.2	0.0006M NaHCO$_3$	20.4	1.2
Methacrylate	(3 to 4)?	0.2–0.6	0.2	0.0015M NaHCO$_3$	NR	NR
Pyruvate	3.5	0.46	0.2	0.0006M NaHCO$_3$	16.1	0.7
Levulinate	(2.8)?	0.17	0.3	None	30.9	2.2
3-Hydroxybutyrate	NR[b]	NR	NR	0.0006M NaHCO$_3$	22.6	1.4
Hydroxyisovalerate	NR	NR	NR	0.0006M NaHCO$_3$	25.4	1.7
Hydroxy-n-valerate	NR	NR	NR	0.0006M NaHCO$_3$	27.6	1.9
Methylethylphosphonate	2.5	0.07	0.2	0.001M NaOH; 0.005M Na$_2$B$_4$O$_7$	23.4	1.5
Isopropylmethylphosphonate	2.4	0.04	0.1	0.005M Na$_2$B$_4$O$_7$	26.9	1.8
Borate/tetraborate	3–4	0.2–0.6	250	ICE/Special eluent	17.6	0.85(C)
Bicarbonate/carbonate	3–4	0.2–0.6	50	IE	25	1.6
Sulfide	4	0.7	250	None	NR	High(?)

[a] Relative retention $= K' = \dfrac{T_{peak} - T_{void\ volume}}{T_{void\ volume}}$ (T = time).

NR = not reported.

(A) Relative to sulfate retention (i.e., K' SO$_4^{2-}$ = 1.0).

(B) Relative to acetate retention (i.e., K' acetate = 1.0).

(C) Eluent: 0.01M HCl plus 0.1M Mannitol.

[b] NR = not reported.

Table A.9. Elution of Moderately Held Anions with Standard Anion IC and Anion ICE Operating Conditions

Anion	IC Conditions				ICE Conditions	
	Approximate Elution Time (min)	Relative Retention (A)	Estimated MDL (ppm)	Other Reported Useful IC Eluents	Approximate Elution Time (min)	Relative Retentive (B)
Nitrite	5.0	1.1	0.5–1	$0.0015M$ $NaHCO_3$	NR	1.3
Azide	9.0	2.7	0.1	$0.002M$ NaOH/ $0.0024M$ Na_2CO_3	NR	NR
Nitrate	12.5	4.2	0.2	None	9.5	0
Sulfite	13.0	4.4	0.3	$0.005M$ NaOH/ $0.004M$ Na_2CO_3	9.5	0
Phosphite	7.0	1.9	0.2	None	NR	NR
Phosphate	8.5	2.5	0.2	$0.005M$ NaOH/ $0.004M$ Na_2CO_3	9.5	0
Monofluorophosphate	6.0	1.5	0.2	None	NR	NR
Butylphosphate	(7.0)?	1.9	0.1	$0.003M$ $NaCO_3$/ $0.001M$ NaOH	NR	NR

Dichloroacetate	7.0	1.9	0.2	None	NR	NR
Benzoate	7.0	1.9	0.3–5(C)	None	NR	NR
Succinate	9	2.7	0.5	None	20.9	1.2
Bromide	10.5	3.4	0.5	None	9.5	0
Malonate	11.5	3.8	0.5	None	18.1	0.9
Maleate	15.0	5.3	0.5	None	14.2	0.5
Tartrate	16.0	5.7	0.5	None	16.2	0.7
Propionate	6.7	1.8	0.3	0.0015M NaHCO$_3$	27.6	1.9
Sulfate	19.0	6.9	0.1	None	9.5	0
Itaconate	15.5	5.5	(0.5)?	None	NR	NR
Butyrate	(6.5)?	1.7	(0.3)?	None		
Malate	NR	NR	NR	None	18.1	0.9
Quinate	NR	NR	NR	None	16.2	0.7
Glutarate	14.5	5.1	0.3	None	25.6	1.7
Methylphosphonate	5.5	1.3	1.3	0.01M NaOH	25.6	1.7

(A) Relative to sulfate retention (i.e., K' SO$_4^{2-}$ = 1.0).
(B) Relative to acetate retention (i.e., K' acetate = 1.0).
(C) Depends on size of suppressor used.

Table A.10. Elution of Strongly Held Anions Requiring Stronger, Nonstandard Anion IC Operating Conditions

Anion	Eluents Used	Separator Length (mm)	Approximate Elution Time (min)	Estimated MDL (ppm)
Iodide	0.006M Na$_2$CO$_3$/0.0075M p-CNP	3 × 250[a]	8	0.5
Thiosulfate	0.006M Na$_2$CO$_3$/0.0075M p-CNP	3 × 250[a]	8	0.5
Thiocyanate	0.006M Na$_2$CO$_3$/0.0075M p-CNP	3 × 250[a]	11	0.5
Tetrafluoroborate		3 × 250[a]		
Arsenate	0.0035M Na$_2$CO$_3$/0.0026M NaOH	3 × 250	24	0.5
Chromate	0.006M Na$_2$CO$_3$/0.001M p-CNP	3 × 150[a]	(7–10)?	
Tungstate	0.008M Na$_2$CO$_3$	3 × 150	(10–12)?	20–50
Rhenate	0.01M Na$_2$CO$_3$	3 × 50	(4–6)?	(50)?
Perchlorate	0.005M NaI/silver suppression	3 × 150	5	
Pyrophosphate	0.01M Sodium citrate	3 × 250	7	20
Tripolyphosphate	0.01M Sodium citrate	3 × 250	12	100
Dithionate	0.01M Na$_2$CO$_3$ or 0.005M NaI	3 × 150	(10)?	2–5
Ascorbate	0.0096M Na$_2$CO$_3$/0.002M NaOH	3 × 150	23	(5–10)?
Trichloroacetate	0.0096M Na$_2$CO$_3$/0.002M NaOH	3 × 150	26	(2)?
Fumarate	0.0096M Na$_2$CO$_3$/0.002M NaOH	3 × 150	33	(1)?
Dioxytartrate	0.0096M Na$_2$CO$_3$/0.002M NaOH	3 × 150	19	0.5?
Phthalate	0.0096M Na$_2$CO$_3$/0.002M NaOH	3 × 150	10	NR
Oxalate	0.0024M Na$_2$CO$_3$/0.003M NaHCO$_3$	3 × 500	23	0.5
	ICE standard conditions	9 × 250	15	(0.5–1)?
Citrate	0.01M Na$_2$CO$_3$	3 × 150	8–10	1

[a] Use of para-cyanophenol in the eluent should allow use of conventional separator columns.
p-CNP = para-cyanophenol.
NR = not reported.

Table A.11. Anions Reportedly Eluted by Anion IC or Anion ICE for Which No Published Elution Information Exists

Hydroxycitrate	Hippurate
Chloropropylsulfonate	Sarcosinate
Bromopropylsulfonate	Trifluoromethanesulfonate
Dithiocarbamate	Dibutoxyethylphthalate
Gluconate	Thiolactate
Selenate	Thioglycolate
Periodate	Thioacetate
Butylphosphonate	

Table A.12. Organic Acids Eluted with Ion Moderated Partition in Contents Analysis Column[a]

	k'	pK_a
Glyceric	0.040	3.55
Oxalic	0.067	1.25, 4.14
Maleic	0.280	1.93, 6.58
Oxaloacetic	0.333	
α-Ketoglutaric	0.360	
Citric	0.467	3.09, 4.75, 5.41
Isocitric	0.467	
Gluconic	0.520	3.86
Pyruvic	0.520	2.50
Glucuronic	0.547	
Tartaric	0.547	2.89, 4.16
Glyoxylic	0.733	3.32
Malonic	0.760	2.83, 5.70
Trans-aconitic	0.813	2.80, 4.46
Malic	0.813	3.40, 5.2
D galacturonic	0.867	
Succinic	1.240	
Glycolic	1.240	3.82
Formic	1.507	3.77
Lactic	1.533	3.86
Fumaric	1.533	3.03, 4.54
Glutaric	1.587	4.34, 5.42
Acetic	1.800	4.76
Adipic	2.067	4.42, 5.41
Levulinic	2.093	
L-pyrrolidine 5-carboxylic acid	2.120	
Propionic	2.280	

[a] Courtesy Bio-Rad, Richmond, California.

Table A.13. Elution of Weakly and Moderately Held Cations with Standard Cation IC Conditions

Cation	Normalized Elution Time	Approximate MDL (ppm)
Lithium	8	0.05
Sodium	11	0.05
Potassium	18	0.1
Rubidium	22	0.5
Ammonium	15	0.1
Methylamine	19	1
Dimethylamine	24	1
Trimethylamine	30	1
Tetramethylammonium bromide	35	NR
Ethylamine	21	1
Diethylamine	33	1
Triethylamine	42	1
n-Butylamine	(38)?	1
Ethanolamine	15	NR
Diethanolamine	15	NR
Triethanolamine	18	NR
Isopropanolamine	16	NR
Di-isopropylamine	16	NR
Tri-isopropanolamine	17	NR
Guanadine	(9)?	NR
Nitroguanadine	(10)?	NR

NR = not reported.

Table A.14. Elution of Strongly Held Cations

Cation	Elution Time (min)	Approximate MDL (ppm)
Magnesium	3	0.1
Calcium	5	0.2
Strontium	7	0.5
Barium	11	1
Tetraethylammonium bromide	NR	NR
Tri-*n*-butylamine	NR	NR
Tetra-*n*-butylammonium bromide	NR	NR
Cyclohexylamine	NR	NR

NR = not reported.

Table A.15. Ions Not Yet Analyzable by IC

1. Isocyanate derivatives
2. Silicate
3. Large phosphonates
4. Nitrilotriacetic acid
5. Ethylenediaminetetracetic acid
6. Ethylenediamine
7. Toluenediamine

Index